CANADARM AND COLLABORATION

ELIZABETH HOWELL

CANADARM <u>AND</u>
COLLABORATION

HOW CANADA'S

ASTRONAUTS

AND SPACE ROBOTS

EXPLORE NEW WORLDS

FOREWORD BY ASTRONAUT DAVE WILLIAMS, M.D.

LIBRARY AND ARCHIVES CANADA CATALOGUING
IN PUBLICATION

Title: Canadarm and collaboration : how
Canada's astronauts and space robots explore
new worlds / Elizabeth Howell ; foreword by
Astronaut Dave Williams, M.D.

Names: Howell, Elizabeth, 1983– author. |
Williams, Dave (Dafydd Rhys), 1954– writer
of foreword.

Description: Includes bibliographical
references.

Identifiers: Canadiana (print) 20200264966
Canadiana (ebook) 20200265873

ISBN 978-1-77041-442-6 (softcover)
ISBN 978-1-77305-628-9 (PDF)
ISBN 978-1-77305-627-2 (EPUB)

Subjects: LCSH: Canadian Space Agency—
Officials and employees—Interviews. | LCSH:
Astronauts—Canada—Interviews. | LCSH:
Outer space—Exploration—Canada. | LCSH:
Astronautics and state—Canada. | LCSH:
Astronautics—Canada. | LCGFT: Interviews.

Classification: LCC TL789.85.A1 H69 2020
DDC 629.450092/271—dc23

Copyright © Elizabeth Howell, 2020

Published by ECW Press
665 Gerrard Street East
Toronto, Ontario, Canada M4M 1Y2
416-694-3348 / info@ecwpress.com

Cover design: Michel Vrana
Cover photograph: "S114-E-6642" (NASA), August 3,
2005. Public domain.

The publication of *Canadarm and Collaboration* is funded in part by the Government of Canada. *Ce livre est
financé en partie par le gouvernement du Canada.* We acknowledge the contribution of the Government of
Ontario through the Ontario Book Publishing Tax Credit, and through Ontario Creates for the marketing of
this book.

PRINTED AND BOUND IN CANADA

PRINTING: FRIESENS 5 4 3 2 1

MIX
Paper from
responsible sources
FSC
www.fsc.org   FSC® C016245

CONTENTS

Foreword by Dave Williams, M.D.   vii

Prologue   xi

Chapter 1   "That's $600 a litre in space paint"   1

Chapter 2   "Three strong Canadian arms on board"   27

Chapter 3   Space cards   46

Chapter 4   Glass ceilings and North stars   71

Chapter 5   Canadarm, cuff links and collaboration   93

Chapter 6   More than just visitors   121

Chapter 7   Nine years an astronaut   142

Chapter 8   "Don't let go, Canada"   166

Epilogue   187

Author's Note   201

Acknowledgements   203

Endnotes   207

FOREWORD

by Dave Williams, M.D.

In the distance was the beautiful blue oasis of the planet Earth cast against the black infinite void of space. It was a moment of a lifetime, riding on the end of an icon of Canadian technology, the Canadarm2, with the Canadian flag on my shoulder. I felt immense pride in being able to follow in the footsteps of past Canadian space pioneers on a path to space that would be pursued by the next generation of Canadians in space. There is no question that Canada is a major spacefaring nation. After the spacewalk, one of the crew floated over to share a thought: "Dave, we in the international program truly understand the space station is just the base for the Canadarm." The wry humour brought a smile to my face.

Exploration is a central part of Canadian history, and certainly a passion to push boundaries in aerospace is part of our legacy. As director of the NASA Johnson Space Center's Space and Life Sciences Directorate, I had a shelf in my office called the "making the impossible possible" shelf. On it there were two items. A model of the Avro Arrow was featured prominently. It sat in the centre of the shelf beside a photograph from one of my team members, David McKay, that showed an image obtained by a scanning electron micrograph from a Martian meteorite found in the Antarctic. The black-and-white image revealed chain-like structures that resembled bacterial organisms, suggesting that life may have once existed on another planet in our solar system. The image and associated scientific data caused considerable controversy in the scientific community; for me, what that image represented warranted inclusion on the shelf. The Arrow model had to be there. It would have stood alone on the shelf were it not for the amazing photograph. Many speculate on what would have happened if the Arrow had transitioned to full operational status. The photo made everyone who saw it speculate on another topic: one of the most fundamental remaining scientific questions, whether life exists or has existed elsewhere in our solar system.

The Arrow is remarkable for many reasons, notwithstanding the fact that the cancellation of the project resulted in a group of very talented Canadian aerospace engineers leaving Canada to help kick-start NASA's first human spaceflights. They all played critical roles that enabled the agency to achieve the goal of sending humans to the moon and back by the end of the decade. The third country to send a satellite to space, Canada has been part of space exploration from the beginning.

Many have dreamed of the possibility of human space exploration. The visionary Russian space pioneer Konstantin Tsiolkovsky once said, "The Earth is the cradle of humanity, but mankind cannot stay in the cradle forever." He and Robert Goddard are widely attributed with describing the principles of modern rocketry that underlie human space travel. It is not so widely known that William Leitch, the fifth principal of Queen's University, described the principles of rocketry and human spaceflight in an 1861 article entitled "A Journey Through Space." Four years before Jules Verne wrote *From the Earth to the Moon*, six years before Confederation and nearly 40 years before Tsiolkovsky's and Goddard's work, a

Canadian scientist had described modern space travel. "We might, with such a machine, transcend the boundaries of our globe, and visit other orbs." Had this discovery been made before 1998, I would have had Leitch's photograph on my shelf as well, but it wasn't until 2015 that Canadian space historian Robert Godwin would discover the original publication. It is clear, though, that Canada is well deserving of the title "major spacefaring nation."

Occasionally Canadians can be understated in acknowledging the global impact of Canadian contributions to space exploration. Yet ours is a story to celebrate. It is a story of visionary scientific and engineering teamwork, a story of pushing the edge of the envelope by incredibly talented scientists, aerospace engineers, researchers, physicians and astronauts. It is a story that I am proud to have dreamed about, studied and lived as a physician, Canadian astronaut, scientist and senior executive at NASA. It is a story that continues to unfold as we set sights beyond the International Space Station in low Earth orbit to the Moon and ultimately to Mars. It is the story of humans pursuing their destiny as a spacefaring species, reaching out to other destinations in our solar system.

Elizabeth Howell is one of the few Canadian scientific journalists to write extensively about space exploration. She brings a unique perspective and true passion to sharing our history exploring space. It's all there. From the Arrow program through Mercury, Gemini and Apollo, the legacy of the Canadian space pioneers captures the reader's attention. The early days of the Canadarm leading to the hiring of the first group of astronauts solidified our international reputation for excellence in space robotics and human space exploration. Describing the continued evolution through the shuttle program years to the era of the International Space Station with multiple astronaut flights, complex robotic operations, technically demanding research missions and Canadians in key leadership roles at NASA, *Canadarm and Collaboration: How Canada's Astronauts and Space Robots Explore New Worlds* gives Canadians fantastic insights into programs that are part of our history as well as the foundation for our international reputation as a leader in the utilization and exploration of space. It is a book that I couldn't put down — in part for the memories that resurfaced while reading it, but primarily for the pride I felt to be one small part of the amazing group of Canadians who made the impossible possible. For anyone interested in space exploration, it is a must-read.

David R. Williams OC OOnt MD CM
Canadian Astronaut STS-90, STS-118
Bestselling author of *Defying Limits*.

PROLOGUE

Sing to me of the man, Muse, the man of twists and
turns, driven time and again off course.

— Homer, *The Odyssey*
(translated by Robert Fagles, 1996)

Nightmares about a rocket abort woke me up early on a cold
Kazakhstan morning in December.

I clearly saw the scenario on this historic launch day:
Canadian astronaut David Saint-Jacques saluting the Russian
commission in his spacesuit, moments after saying goodbye to
his three small children and physician wife. His cold bus ride
to the launch pad at Baikonur, sitting alongside his two crewmates. A final wave to colleagues and friends gathered at the
foot of the rocket.

A flawless liftoff to the International Space Station, marred
by a sudden vibration. Saint-Jacques, in the left-hand pilot's
seat, would scan the abnormal readouts on the console. Either
he — or Russian commander Oleg Kononenko — would speak
with Russia's mission control calmly, scientifically, even as their
abort system pulled them violently towards Earth.

"Two minutes and 45 seconds," they would say in Russian as they read the elapsed time since their rocket lifted off. They would report "the emergency, the failure of the booster" as the abort system took over. And they would keep astronaut-ing all the way down to the ground, doing everything possible to keep the spacecraft from crashing into the Kazakh Steppe.

It wasn't just conjecture haunting my dreams; this is what had happened to an American and a Russian during an abort less than two months before, on Oct. 11, 2018. They arrived back safely, and the abort system worked flawlessly, but it was scary as hell to watch on NASA Television.[1]

Here's how NASA astronaut Nick Hague, who occupied the left-hand seat on that failed mission, later described his feelings as alarms blared and the spacecraft suddenly went weightless: "There's a little bit of disbelief," he told NASA. "It hasn't happened in 35 years, so that was a bit surprising."[2]

I lost sleep over the possibility of rocket failure, yet Hague manifested pure calm when faced with it in real-time. Why?

"My career in the [United States] Air Force has done a lot to help me prepare for stressful situations like this, whether it's through deployments or my time in flight test, where we had to deal with failures in aircraft that you're in and having to get down on the ground immediately," he added. "You really end up falling back to your training . . . Over my two decades in the Air Force, I've learned that in those situations, the best thing that you can do is stay calm and do what you've been trained to do."

After the abort, the Russians had put together a group of experts, the State Commission, which is usually chaired by a very senior-level Russian space agency (Roscosmos) manager. "They'll bring in all of the relevant entities and evaluate what went wrong," said Chad Rowe, NASA's director of human spaceflight programs in Russia, of the typical role of the State Commission. Rowe helps coordinate much of the day-to-day work of NASA employees in Moscow and Kazakhstan, assisted by an able group of interpreters when they interface with Russian technical staff.[3] Rowe also helps with the logistics for media and other dignitaries who attend launches, including Saint-Jacques's.

"In October, with that launch abort, they had a really good idea about what had gone wrong, the night of the launch," he explained. "They shared some of the information with us, and they said what their next steps were going to be. They went off and did this analysis to validate that their assumptions were good. Once they did that, they went through a large number of evaluations for how to mitigate the likelihood of recurrence in the future."

Meanwhile, Rowe's boss, International Space Station (ISS) program manager Kirk Shireman, set up his own group of experts for an independent investigation. "We take whatever we can learn from what we know, and then bring in our own experts and start to suppose what probably went wrong, and what went right, what we would do as a technical organization to mitigate recurrence," Rowe said.

"While we're [NASA] waiting for initial responses and conclusions from the State Commission, then we really are able to hit the ground running because our experts have gone through relevant thought processes and literally have pages of thoughts as to what probably went wrong and what could be done."

NASA and Russia thoroughly investigated the issue and found the faulty rocket sensor that stopped the Expedition 57 flight short of its goal — which was a huge success story. Rowe pointed out several things underplayed in the media: that the abort system worked perfectly, that both Russia and the United States came carefully yet swiftly to the same conclusion about the mission failure, and that both sides had the benefit of a lot of experience to help them learn how to address the problem. After all, the Russians have been launching one variant or another of the Soyuz spacecraft since the 1960s. Not to mention, the two countries have been working together for so long — a generation — that they very much trust each other's expertise.

"The Russian team here . . . they have very unique and specific capabilities across the whole gamut of space. They're the ones that are still bringing all of our astronauts to the space station right now. They have a highly reliable launch capability," Rowe added, praising the launch vehicles and spacecraft for their simplicity, robustness and reliability. In fact, he likes simple: "The more complicated things are, the more things that can go wrong. Consequences from any of those things that can go wrong in space are much higher than the consequences of engineering projects going wrong on Earth."

I knew this stuff intellectually. I knew Soyuz was amazingly reliable, even when things go wrong — after all, we had just seen two people come unscathed through an abort.

But it's one thing to think about this impartially. It's another to watch a Canadian — a Canadian I know fairly well after 10 years of reporting on his activities — climb into the beast and dare to ride atop its flames.

Besides which, so much on this trip had already gone wrong.

It takes $5,000 and special permission from the Canadian Space Agency, NASA and the Russian authorities to get approval to watch a Canadian go to space from Kazakhstan. There are visas and insurance forms to fill out, security briefings to attend and endless packing lists to follow.

The morning before I was supposed to leave from Ottawa, reports of an impending snowstorm forced me to change tickets and pack in 20 minutes to catch an early flight. It took me two days and many flights and transfers and airports to get to my first destination, Moscow's Vnukovo Airport.

Upon arrival, I discovered the international customs line wasn't a line at all; it was a restless mass of people facing some sort of undisclosed delay at passport control. Worse, I had a NASA driver waiting for me somewhere near the airport entrance and no way to reach him; after talking with friends about computer security in Russia, I had left my cellphone behind out of fear of it contracting a virus.

I waited for 45 minutes, listening to complaints in the crowd. At one point, Russian custom officials discovered a few people were in the wrong line. "*Oohadee!*" the crowd yelled, encouraging those unfortunate folks to leave and get in the right passport control zone. But even with them gone, I and my heavy backpack barely budged.

At last I resorted to going to the empty diplomatic line to try my luck. I handed over my Russian invitation visa. "*Oo menya predlojenia Roscosmosa,*" I told the passport official in my poor Russian — "I have an invitation from Roscosmos," the Russian space agency. I expected to be questioned and selected for secondary screening, but the officer barely looked up from his paperwork. "What was your flight number?" he responded in perfect English. Processing me took less than a minute. I was in.

I met up with my group at the Volga Hotel, got a prelaunch briefing, and about 36 hours later, found myself back at the same airport facing a flight delay. Now there were about 50 of us — Canadian journalists, NASA officials and the Canadian delegation to see Saint-Jacques's launch. Something mechanical was holding up our plane.

We left for Kazakhstan two hours late on our charter flight, and all appeared to go well until we suddenly turned around in mid-air. Turns out that our delayed departure meant we would land at Baikonur in darkness, which was a problem because there are no lights at the civilian Baikonur airport. After turning around in mid-air, we returned to Russian soil, to a place called Samara. Our limited-entry visas wouldn't permit us to easily exit the airplane, be processed again in Russian customs and go on to Kazakhstan, so NASA swiftly negotiated for us to take off again and be rerouted to Kyzylorda, Kazakhstan — a three-hour bus ride from Baikonur.

We finally arrived in Baikonur after an unexpected over-night, overland journey — most of us on very little sleep. There was barely time for a nap, as our first event was in only two hours. I rested in my hotel room and left with a few extra clothes on to watch Saint-Jacques's rocket emerge from its hangar at 7 a.m. sharp, amid a bitter wind that we Canadians estimated dropped the temperature to at least −20 degrees Celsius. I did some iPhone filming for another Canadian journalist; my bare fingers froze to the metal within seconds.

NASA and Roscosmos kindly provided our contingent of Canadian journalists with a driver, a fixer, a translator and a van to herd us from place to place in Baikonur, a military base with limited access for the public. Our second stop was a railway crossing, where we could watch the rocket go through. I was too exhausted and cold to venture from the van to the crossing, some 500 metres away, but I stepped outside to watch the rocket-beast flow through.

When it was time to get back inside the van, I hauled on the passenger door, but it refused to open. Two people tried operating it from the inside and rapidly realized they were trapped. The van driver hopped over his seat and pulled on that handle every way possible, before taking his toolbox outside and disassembling and reassembling the door mechanism in the middle of the dark field. Underdressed and shivering, I finally climbed back into my seat and basked in the van's heat.

So yeah, on the little sleep I had at the time, after customs lineups, delayed flights and broken door handles, a rocket abort seemed very possible. In retrospect, these problems were small, and I should have followed the advice of Canadian astronaut Jennifer (Jenni) Sidey-Gibbons.

"One of the things that really kept me going," she told me of her many tests in fire and water and with limited time conditions during tests associated with astronaut selection, "is that, if I take it easy on this task, or I don't do my best, or I don't push harder, it's going to make it more difficult for my team to complete the problem, and it's going to be harder for my teammates."[4]

Perhaps if I had treated my fellow reporters as teammates and sought to help them, together we could have made this more of an adventure and less of a problem to be solved.

To be clear: I was so grateful to be in Kazakhstan, fulfilling a lifelong dream to see a Canadian launch to space. But the logistics felt terrible and exhausting after only three days working on the story. Would Saint-Jacques have any more luck reaching the space station?

We only had limited access to Saint-Jacques. He was, necessarily, in quarantine and saying his goodbyes to family, but he assured us in two press conferences that he felt ready to go. I asked him how his family and friends were supporting him on this long training journey, which had already taken Saint-Jacques more than two years to accomplish even before he left for six months in space.

"We are going to keep doing the very successful recipe that we found, which involves everyone we love lending a hand and helping us," he told me. "This is really huge teamwork on the home front, from friends, family. I'm glad, because the fact that my family, the people I love most, are in such a good spot, really, allows me to leave with peace of mind."[5]

I didn't feel peace of mind on Dec. 3 — launch day — but Saint-Jacques appeared serene. In the suit-up room, separated from a crowd of nosy journalists and onlookers by only a glass window, he spent 30 minutes calmly speaking with his three children. They talked by microphone, and he allowed them to discuss whatever came to their minds — even their complaints that he wouldn't be around for ski season this year.

Russian officials eventually told us to clear the room. Saint-Jacques asked for his family to come close to him. *"Les enfants, les enfants!"* cried the francophones in the crowd, and the children ran to the window, pressing their faces and hands against the glass. NASA astronaut Anne McClain, a mother herself, stood watching them with a look of love and sympathy. I was crying, but rocket schedules don't wait for emotions.

Outside, I negotiated (as much as I could with my limited Russian) with one of the many armed military officers in the area. I wanted to stand close to where the crew would say goodbye to us Earthlings. These officers, by the way, were lovely with journalists — even though they spoke no English and few of us spoke Russian, they did what they could to support us in getting our footage.

He gently challenged me: *"Kto vee?"* — who are you? — but when I responded, *"Canadski press,"* he allowed that it would be okay for a Canadian journalist to go stand over by the TV cameras. Luckily, I got a superb location. I saw the astronauts walk from the building, oxygen tanks in hand, and stood less than six feet from Saint-Jacques when fellow crew member Oleg Kononenko announced they were ready to carry out their duties from space.

Can you imagine a Russian-led launch like this in 1984, when Marc Garneau became the first Canadian in space aboard an American space shuttle? Canada's involvement in space went from our astronauts managing a few small experiments on a space shuttle back then to, in 2012, managing an entire space station.

Our astronauts have spacewalked, operated the Canadian robotic arm, and — twice! — flown the pilot's seat in Russian spacecraft. Piloting Russian craft would have been difficult to imagine a few decades ago because even the Americans would not let foreigners near the flight controls of their space shuttle.

But it was through Canada's ISS journey, and our acceptance of a non-traditional space partner in Russia, that we got to where we are today.

This story is about how Canada persisted.

CHAPTER 1

"That's $600 a litre in space paint"

Ground control to Major Tom
Lock your Soyuz hatch and put your helmet on.
— Chris Hadfield, "Space Oddity"

A miffed member of Parliament wasn't going to let a little security stand between him and the first man on the moon.

Conservative MP Jack Horner reportedly dodged past guards at Parliament Hill and slipped into a secure cabinet meeting when he heard Neil Armstrong and his Apollo 11 crew were nearby. When he crashed the Dec. 3, 1969, meeting, a sympathetic secretary of state, Paul Martin Sr., kept up appearances by bestowing the intruder with an honourary title to waylay curious glances: "Senator Horner."

For Horner, it hadn't been enough to receive an invitation to a swanky dinner with the Apollo 11 astronauts that night at Ottawa's Château Laurier (along with 500 other honoured guests). He interrupted the proceedings on Parliament Hill to slip a rock from Alberta into Armstrong's hand.

"You've travelled all over heaven and Earth picking up rocks," he told the astronaut, "so here is one from another forgotten place — the prairies."[1]

Horner was far from alone in his enthusiasm. The unfazed "moonmen"[2] (or if you prefer the francophone journalists' term, *lunautes*[3]) faced down several more security issues during their daylong tour of Ottawa. At Centre Block, frustrated children rushed under a security rope and mobbed Armstrong, Buzz Aldrin, Michael Collins — and even Prime Minister Pierre Trudeau.[4] At dinner, at least six members of Parliament interrupted the astronauts as they ate, hawking autographs; one journalist witnessing the sight lamented the "astronaut fever . . . which reduces $18,000-a-year MPs to shoving, elbowing autograph hunters."[5]

The astronauts found themselves facing some profoundly Earth-bound queries. One reporter asked Collins what he thought of the US Army slaughtering hundreds of unarmed civilians in the My Lai Massacre, the notorious Vietnam War fiasco that had occurred a year earlier. Collins was then nominated as assistant secretary of state for public affairs, and the matter was still making headlines in late 1969, but the question induced resounding boos from the press gallery. (Although My Lai happened in 1968, it didn't become public knowledge until November 1969, a couple of weeks before the Apollo 11 visit to Ottawa.) Collins hastily replied that the allegations were "tragic, if true" and refused to go any further.[6]

While protocol breaches dominated the headlines, some journalists wondered just why the astronauts were in Ottawa, and later Montreal, in the first place. Officially, Canada was the astronauts' last stop on their months-long goodwill tour after their July 1969 moon landing. The astronaut spoke only briefly to the public outdoors, understandingly giving the excuse that Canada is cold in early December.[7]

Enough angry letters came to Prime Minister Trudeau's office afterwards that an anonymous official created a boilerplate response: "The guests made their wishes known," the letter said in part, that they were not prepared to follow Canada's original agenda of public events — which included meeting with children. "[I]t was obviously the duty of Canada as host to think of the visitors first and of our own preferences second."[8]

In fact, the visit was a nagging reminder that Canada's own space program appeared "grounded," if you were to believe one headline from the *Edmonton Journal*. According to that report, a major review of Canada's space activities practically begged the country to set up a central space agency, but the government had not budged. Canada had been firing rockets from the shores of Hudson Bay at Churchill since 1954 to tease out the secrets of the northern lights and other atmospheric phenomena. But that successful program was in turmoil after major partner NASA left and the National Research Council (NRC) slashed its contribution on austerity grounds.

Telesat was created in 1969 as a Crown corporation to manage satellite operations that other branches of the federal government had been managing since Canada became the third country in space with the launch of Alouette in 1962. In the ensuing decades, Telesat was privatized. The emergence of a commercial competitor for this responsibility was rubbing a lot of government people the wrong way, especially those with vested interests. Telesat found itself in a turf war, with the Department of Communications and the Department of Energy, Mines and Resources fighting over who was the true owner.[9]

"For better or worse, Canada now has a space program determined not by a central agency with its finger on every aspect of the problem, but by aggressive cabinet ministers who grabbed what was available,"[10] concluded Peter Calamai, who just two years after that 1969 visit co-founded the Canadian Science Writers' Association (today's Science Writers and Communicators of Canada, of which this author was president in 2019).

3

So maybe it didn't matter when the astronauts thanked Montreal company Héroux for manufacturing their lunar lander legs,[11] or when Aldrin said in a speech that several nations might someday build a moon base together.[12] Maybe Canada was a tiny country burdened by other priorities and needed to leave space to bigger players.

It is easy to think that, looking at where Canada was in 1969 — no space agency, no astronauts, only a scattered industry necessarily focused on telecommunications technology to unite our large country or understanding the northern lights and the upper atmosphere to improve radio communications.

But here's the truth: Canada was mightier than many knew. That same year, NASA spoke with Canada about international participation in developing its new space vehicle, something called the space shuttle. In contrast to the needle-nosed rockets that had been sending satellites and people into space for more than a decade, this spacecraft would look more like a conventional airplane, complete with wings and a tail rudder. Nor was it to be a one-time-use vehicle like the space capsules, which were only good for museums after returning to Earth. The space shuttle was intended to fly again and again, serving as a sort of cargo truck designed to bring satellites and large space hardware into orbit.[13]

The space shuttle, like NASA's other initiatives, was to be very much an American undertaking. But within a few years, it offered Canada a rich opportunity in the form of spinoff technology from an antenna — yes, an antenna! — that would earn this country a front seat in NASA's astronaut program. Maybe you've heard of this spinoff, called Canadarm.

Seven years before the first human being walked on the moon, Canada made its own mark in space history with Alouette, the remarkable satellite that flew on Sept. 29, 1962 — making Canada only the third country to touch space after the United

States and Russia. Engineer Colin Franklin — born in New Zealand, migrant to Britain and eventually Canada — remembers watching Sputnik shine in the night sky in 1957, but he never could have imagined that starry messenger would lead his Ottawa-based Canadian group to close collaboration with the Americans five years later.

The Americans were keen on eventual international collaboration, in part to offset launch costs and in part to show they had a more open society than the Soviets. Eldin Warren, a scientist at Canada's Defence Research Telecommunications Establishment, made a proposal to fly a Canadian experiment on board a US satellite. NASA felt the proposition was too technologically difficult when it was proposed in 1959 — after all, space exploration was in its infancy and even the simplest tech was a huge risk to the mission — but asked the Canadians to keep working at it.

"You can imagine that would just kill the whole thing, just on the spot," Franklin recalled in an interview.[14] "However, it didn't do that with us. And we actually started work on this program, proposal, experiment, at the beginning of 1959. It was actually January 1st or 2nd of 1959. And NASA didn't open its doors [officially to international partners] until, I think, October 1959."

This initiative eventually took physical form as an independent Canadian satellite called Alouette, supported by no more than a simple exchange of letters between the head of NASA and the chair of the Defence Research Board, Franklin said. "Can you imagine that nowadays?" he added.

But at the time, it worked well. Of course, there was no guarantee that the American Thor-Agena rocket would, as agreed, survive long enough to hoist the Canadian Alouette into space; the technology was so young that rockets routinely blew up on the pad. On the other hand, Franklin added, there were no financial constraints in the agreement — if the Canadians wanted more test equipment, there never seemed to be a worry about how much it would cost. Alouette, however, did ruffle feathers when personnel and funds were diverted from other programs to support it.

This young Canadian team — made up largely of people in their 20s and 30s — were working at the very cutting edge of technology, implementing nothing but semiconductors on their 1960s-era satellite. (If that seems brave, know that Franklin himself worked on semiconductors for his Ph.D. at London's Imperial College in 1953, back in an era when even transistors were rare.) The team was criticized for implementing testing procedures such as exposing their semiconductors to extreme temperatures, on the grounds of unnecessary risk to the components. It took the personal support of John H. Chapman, a prominent figure in Canadian space history, to get the testing approved, Franklin recalled, at some risk to Chapman's career. Today, that testing is just standard space procedure: best to know if something will fail before it leaves the launch pad.

In some ways, these talented young men were treated like a bunch of wildlings. A senior manager, Frank Davies, lectured the Alouette team on how to behave at Vandenberg Air Force Base in California, imagining that this team — with no girl-friends or wives on site — might have too much time on their hands. They were even instructed not to shine their shoes with the hotel towels.

Despite the criticism — and the towel advice — the 30-strong group showed up at Vandenberg on two separate planes, each plane containing an identical working satellite and its own team. Vandenberg Air Force Base, the historical US Army base and newly named and reframed rocket complex, was already famed for launches in the early US space program — such as that of the first US satellite, Explorer 1. So, Canada had made it to the big time, and the backup second satellite was something they obviously took very seriously as part of pleasing their advanced American counterparts. Testing proceeded up to the last minute, just to make sure everything was okay. A big problem did come up in the six or seven weeks on site; the team discovered their satellite was too light for the rocket to safely bring into space. "They ballasted it with concrete," Franklin said.

Then there was the problem of making sure the range was clear for launch. The launch was aborted at least twice due to a California fruit company that sent trains between San Francisco and Los Angeles; the tracks, through unlucky chance, ran through the launch site.

"The fruit company, I forget the formal name for their fruit company, absolutely were not interested in telling us when they were coming. And NASA didn't know," Franklin said. So each launch required watching the tracks. "Someone, NASA or Air Force people, would look out and shout out that the fruit train was coming," he added.

National broadcaster CBC, frustrated that (of all things) a fruit train had twice held up this high-tech launch, refused to continue watching the shenanigans and left the site after the second abort. Only one Canadian reporter stuck out the long wait — Lowell Green, a young correspondent at radio station CFRA. By nature, reporters are an impatient bunch and the complexities of space are difficult to grasp for those who may be new to the field, but this chain of events is a reminder that *everyone* was new to the field — even the experts.

When Alouette finally launched, everyone waited anxiously for it to fly over its first communications point — a ship. In contrast to today's extensive global network of relay satellites and ground stations that maintain constant contact with anything in orbit, such links could only be established intermittently, and "dead zones" were common during flights. Unfortunately, the ship never received the signal — and Franklin briefly thought all was lost.

"It turned out that the ship had communications problems. So it didn't receive [the signal]," he recalled. Several more tense minutes of waiting passed until the next communications point at Fairbanks, in Alaska. An engineer technician from the team had flown all the way from Canada to Fairbanks (no small feat in 1962) and was watching a sea of printed-out numbers for one vital thing.

"He reported that the antenna was extended," Franklin said, a key indicator that the satellite was doing very well in its new environment. With the antenna extended, Alouette could receive and transmit information, and the mission was ready to start the vital commissioning period — the few weeks or months where ground engineers test a satellite to make sure it's ready to do its science. Alouette was designed to last for one year, but it exceeded that time frame tenfold. In a decade of observations, it produced a million images of Earth's ionosphere.[15] Learning about the ionosphere is crucial for radio communications and also for figuring out the structure of the Earth's atmosphere, which can change according to factors such as how active the sun is.

And remember that backup Alouette craft, so carefully carried to California on a separate plane with a separate crew? It had its day. Alouette II launched successfully in 1965, continuing the atmospheric studies — and more importantly, Canada's early run in space.

The Canadarm program (although it wasn't called Canadarm back then) launched in 1974 under the able management of Jeanne Sauvé, then minister of state for science and technology, who wanted to build a sustainable industrial framework to support Canada's push into space. Canadarm was Canada's space currency. We used it (and still use space robots like it) to get seats for our Canadian astronauts. It was a 15-metre arm emblazoned with a Canadian logo, painted with ordinary hardware-store paint by engineers wanting to advertise our country in space. The paint cost of $4,500 paled in comparison with NASA's response — putting the American flag on the space shuttle at a cost of $450,000 — at $600 a litre in space paint. "In an interesting test," Canadian space policy analyst John Kirton wryly wrote in 1985, "both have stood up fairly well."[16]

Canadarm's beginnings came from a humble tobacco-filled tube. Canadian inventor and NRC employee George Klein was sailing from Liverpool on the North Atlantic in 1951 when he looked down at the cigarette paper he had rolled countless other times and realized he could make something useful from it.[17]

The swirly tube design eventually spawned a series of antennas known as Storable Tubular Extendible Member (STEM), which incidentally is also a good name for a band. I saw one of the 1960s' versions of the antenna up close, during a tour of the archival building of the Canada Science and Technology Museum in December 2018.[18] It's a heavy tube that looks like an awkwardly shaped fishing rod. Canada is so proud of this STEM that several versions of it lurk in the museum archives, but the two officials accompanying me said the one I saw is a great example of STEM's early appearance. STEM stuffs its antenna into a tiny box for convenient stowage during liftoff, and then it can be "unrolled" into the long antenna shape. IKEA couldn't have packed it better.

STEM flew in space on Canada's first satellite, Alouette 1, in 1962 and quickly found application on the crewed space program as well. Astronauts in all the early programs — Mercury, Gemini, and, yes, the Apollo missions to the moon — depended on this invention. The design is so innovative that NASA persists in using it today. A STEM antenna will fly on the James Webb Space Telescope[19] that will search for exoplanets, distant galaxies and other secrets of the universe.

Canada was just as well poised to help out when NASA began musing about some sort of reusable space vehicle that would make access to space routine and cheap. The impressive STEM track record of De Havilland and its spinoff space company, Spar, inspired Canadians to make a big bet — to go for the remote manipulator system NASA wanted for its shuttle in 1975.

From 1969 to the beginning of 1974, several studies were carried out to determine how Canada might proceed. Beginning in 1974, NRC started funding the industrial team to do feasibility work and started working with NASA. This led to the signing of a Memorandum of Understanding between NRC and NASA in 1975, ratified by an exchange of diplomatic notes in 1976.

Gaining the support of the Canadian government — with no central space agency able to lobby for Canadarm's existence — took determined effort by Minister Sauvé. She lobbied individual members of cabinet to support the $100-million venture.[20] But once the Canadians were on board, it wasn't hard to sell NASA on the idea. Spar and RCA Canada Ltd. (which later also became a part of Spar) were both able space engineering companies, and NASA clearly loved STEM. Spar became the prime contractor on Canadarm and asked Klein himself, then 72 years, old to provide an independent review of the SPAR gear design. He remained active in aerospace activities well into his 80s.[21]

Canadarm is a magical tool reminiscent of science fiction stories about robots in space; during 30 years of orbital visits, it hoisted spacewalking astronauts, rescued stranded satellites and helped to build the International Space Station. It's easy to take engineering feats for granted after they're completed. But making Canadarm work in space was no small matter. The arm couldn't even hold itself up under Earth's gravity, which meant it needed to be tested on special sliders on the ground.[22]

Testing was performed at Ottawa's David Florida Laboratory, originally a Communications Research Centre facility specializing in esoteric work on satellites. Normally, you can park a car in the vacuum chambers available, but Canadarm was too big to fold inside. Testing therefore had to be done on individual joints instead.[23]

This work meant a great deal to Garry Lindberg, a young engineer at the NRC who had graduated with a Ph.D. in engineering mechanics at the University of Cambridge before joining NRC in 1964. While he managed a variety of engineering programs, Canadarm was his most complex project to date. Lindberg, who led the engineering effort, said hundreds of little issues came up. One "controversial" decision about how to snag a satellite in space (as it was not clear this technology would work) resulted in a hallmark of Canadarm technology: the end effector. This circular wire contraption allowed the arm to "snare whatever you were capturing, but not try to do a rigid grasp," he said.[24] But it took a lot of convincing of engineers at NASA's Johnson Space Center (JSC), the home of the agency's human space program, to accept this for flight.

Another issue had to do with gears: "If you look at the arm, you don't want it to wiggle or wriggle, and so you had to design something that [had] zero backlash, which means you didn't want it to be gearing to move backwards," he recalled. "That was a very challenging design effort, and early on, the decision was made to use very large gear ratios and relatively small motors."[25]

The real test of space hardware comes when it's actually used in orbit. NASA likes to list its space technologies by "technology readiness levels" to help its organization understand how worthy a new concept is for spaceflight. The gold standard is an actual space mission. Canadarm's turn came on only the second space shuttle mission, in November 1981. Lindberg, funnily enough, was in a television studio when the Canadarm tests took place.

"I'm sitting in the studio in Toronto, and the orbiter disappears out of earshot," he recalled, meaning the space shuttle was out of reach of any satellites or ground stations. "If the test was going to happen, it would happen in the silence." With a live television camera on his face, Lindberg tried to compose himself for any news.

Meanwhile, in orbit, astronauts Richard "Dick" Truly and Joe Engle were doing their best to focus while on a shortened flight. A faulty fuel cell had cut the planned five-day mission to less than three. NASA was careful not to rush the astronauts — they basically redid the flight schedule to include only the most important tests — but nearly 40 years later, Truly remembered the pressure.

"We were in a scramble to get all the flight tests done," he recalled in an interview, "but we did it. We — Joe and I — unberthed the arm and ran it through all its test procedures. It worked like a champ, and so it was a great success as a test flight."[26]

Once the shuttle astronauts called home with the good news, a relieved Lindberg allowed himself to show his happiness on camera. "I probably gave them a good reaction," he remembered. "But they also would've captured what my face looked like if [the arm] hadn't moved or hadn't functioned."[27]

And that was just the beginning of this robotic arm's career, which eventually saw it used alongside astronauts — and even carrying astronauts — during spacewalks. Brampton-based company MDA bought the Canadarm technology decades ago with its purchase of the Canadian company called Spar Aerospace, which expanded its expertise in space robotics. The emerging era of "collaborative robotics" in the 1980s is something we take for granted today. Canadarm was the most high-profile example of fusing human and machine in space, to an extent that only science fiction writers dreamed of in the decades before.

"The idea that a person even got near a robot really [gave] the concern that they'd actually be hurt by that robot," said

Cameron Ower, director of engineering and the chief technology officer for robotics and automation at MDA.[28] Canadarm, he said, "originally conceived as this large crane-like manipulator to move a payload like a spacecraft out of the shuttle bay for release into orbit — or do the opposite, retrieve it and bring it in for repair. [But] because of the way that Canadarm had been developed, it became apparent that the system could be safe for an astronaut to work beside Canadarm, or even actually ride on the end of Canadarm."

He said some of the greatest examples of Canadarm's use came when astronauts would ride on the end of the robot and do more "dextrous" tasks, while being supported by the arm. Perhaps the most famous example of this work is the multiple repairs and upgrades to the Hubble Space Telescope after its launch in 1990, which allowed the telescope to receive a vital modification from its initial myopia — an aberration in its mirrors that initially blurred early pictures of space. That first launch to repair the telescope in 1993, followed by five more launches through the remainder of the shuttle program, upgraded instruments and swapped out failed parts. The last set of repairs in 2009 was supposed to hold for about five years, but this last repair and upgrade mission was so flawlessly executed that the telescope will likely last well into the 2020s.

Canadian robotics also looked at the integrity of the space shuttle and the International Space Station. (Canadarm's inspection of the shuttle was accomplished using the Orbiter Boom Sensor System, which was also developed in Canada — technology that allowed the space shuttle to return to flight.) "We never thought having [a] roving eye-in-the-sky of the camera would be so crucial," said MDA's Mike Hiltz, who is space station program manager of the systems engineering group and real-time support.[29] "We're able to monitor the vehicle's performance. You can go look at things like stuck or frozen thrusters, waste and water dumps to make sure the vehicle is performing properly. You just couldn't do that if you didn't have a robotic system. Canadarm1 and Canadarm2 were used for that kind of operation."

Canadarm, Canadarm2, and Dextre robotic hand (which will be covered later in the book) are all name-brand projects for MDA. This has helped with their exports, MDA employees said, and allowed them to do other space projects, such as creating end effectors for Japanese space robotics. Not to mention, MDA has extended the work of these robotics well beyond space exploration. Some of their applications are medical — like neuroArm, a surgical system that can perform precise neurosurgery, while integrating real-time MRI, under a trained surgeon's hand.

In 2008, neuroArm made history when it assisted Foothills Medical Centre personnel in removing a tumour from then-21-year-old Paige Nickason's brain. "I had to have the tumour removed anyway, so I was happy to help by being a part of this historical surgery," Nickason laconically said in a press release.[30] *The Globe and Mail* was more effusive in its praise: neuroArm, it said in an article at the time, "removes the physical constraints of the neurosurgeon to offer a rock-steady hand for precision procedures and superhuman vision to the microscopic level."[31]

"We're transferring the control software to allow the surgeon, using this surgical manipulator as an extension of his hands, to do surgery where you're operating down to incremental motions of the tips of the surgical arms, of the order of 50 microns. We're talking about 1/20th of a millimetre," Ower explained. "So this is really fine motion, all coordinated with the same software that was controlling Dextre's motion."

MDA has extended this technology with a company in Toronto, called Synaptive Medical, which works in port-based neurosurgery. With this type of procedure, the arm acts as a surgical assistant. A port is introduced into the patient's brain — like a tube opening — and all the surgery happens through "this sort of keyhole," Ower explained, to do procedures such as removing a lesion from a patient's brain. The assistive robot, known as Modus V, follows the motions of the surgeon and keeps an exoscope trained on top of the port so that the surgeon can see what they are working on. This clear field of view can in some cases, cut down procedures as long as two 8 hour blocks by half. "So that just shows how robotics can sometimes really improve the pace and maintain the quality of an existing practice."

Many years later, astronaut Dave Williams proposed using STEM as a device to create a temporary translation path for space-walking astronauts — a proposal that never went anywhere, he told me, but could "save a lot of time in getting from one location to another during a spacewalk."

The last Canadarm flew in 2011, along with the last space shuttle. NASA was building newer spacecraft, and it was prepared to focus its efforts on the International Space Station in the meantime. While Canadarm no longer flies, you can see one of the three remaining arms in its home country at the Canada Aviation and Space Museum in Ottawa.

And Canadarm spawned successors in medicine and in spaceflight. The next robot to reach orbit was, appropriately enough, named Canadarm2. It still lives on the space station, and it happens to grace one side of the five-dollar bill. In 2019, Prime Minister Justin Trudeau (son of Pierre) announced a Canadarm3 that would live on NASA's Lunar Gateway, a proposed space station out by the moon. So the technology keeps giving both on Earth and in space.

"It's really turned out to be an icon and a great repre-sentation of Canadian engineering, science and technology capabilities. I think all of us involved feel pretty proud having been part of it," Lindberg said.[32]

But there was a dark side to Canada's space program as well. Before Alouette or Canadarm was in the works, and before Canada had an astronaut program, we lost untold years of experience in a devastating industry event known as "Black Friday." That happened on Feb. 20, 1959, after the Diefenbaker government decided not to pursue development of the Avro CF-105 Arrow, an advanced fighter jet. Looking back, it's lucky that Canada managed to pull together a space program at all.

Arrow was advanced for its day — it was designed as a "supersonic all-weather interceptor aircraft," according to *The Canadian Encyclopedia*. In other words, no matter what weather Canada (or any other country) threw at the pilot, he (for it was always a "he" back then) could swoop towards threatening aircraft at faster than the speed of sound. Prototypes were made, but the brand-new weapons and guidance systems made the program costs climb astronomically.[33]

And the plane quickly seemed dated in the face of the intercontinental ballistic missile (ICBM) threat from the Soviet Union. Its leader, Nikita Khrushchev, declared that the USSR was building these missiles "like sausages" and that bombers were an obsolete technology. While later photos from reconnaissance satellites and U-2 spy planes showed the threats were empty, in the absence of information, Prime Minister John Diefenbaker took the threat seriously. By September 1958, Diefenbaker's cabinet concluded it would run the Arrow program only until March 31, 1959, at which point officials would perform a review. The Arrow's manufacturer, Avro, thought it had more time, but cabinet feelings soured quickly against the Arrow, and on Feb. 17, ministers concluded they would wrap it up.[34]

The sudden decision threw more than 14,000 workers into unemployment,[35] including Bruce Aikenhead, who was only six months into his job at Avro working on flight simulators. While everyone struggled to find work in Canada, the aerospace industry in the United States was ready. It was only two weeks later that several American companies put advertisements in Toronto newspapers.

"I remember meeting everybody downtown in the corridors of hotels. You would see everyone in the corridors. This went on for a week or 10 days . . . You'd see some of the same guys all over again. Guys heading for interviews in Long Island or the West Coast or St. Louis or wherever," said Aikenhead in a September 1995 interview with Canadian journalist and historian Chris Gainor,[36] who generously gave me his notes in late 2018 when Aikenhead's family said the 95-year-old aerospace engineer was too ill to do an interview.*

Aikenhead was mulling over offers when a former Avro supervisor called him. Turned out that Jim Chamberlin wanted to bring a group of Avro people down to work for a brand-new organization called NASA. The 1959 NASA had no astronauts, nor built spacecraft — but it did have a tenuous commitment from President Dwight Eisenhower to proceed with a crewed space program.

"NASA was very interested to hear that hundreds of engineers were available. They said there was no way they could compete with industry for wages, but it was a government organization with influence at the State Department that could expedite things through the embassies and consulates that industry couldn't," Aikenhead said.

Just half of those people made it to the interview stage, Aikenhead among them. Then just two weeks after his NASA interview in an old Avro building, he was in — along with 25 or 30 other people, he recalled. "NASA was as good as its word, and the consulate in Toronto was shut down for everyone except us, and we got the red-carpet treatment," he said. The

---

* Bruce Aikenhead passed away on Aug. 5, 2019 at the age of 95. SpaceQ story — https://spaceq.ca/canadian-space-pioneer-bruce-aikenhead-passes-away/

news must have been a relief to Aikenhead, who had four children; the youngest was only four months old.

In 1959, NASA was working on a program called Mercury, which was dedicated to testing how the human body fared in space travel and — in light of the recent Soviet success — making sure the body in question belonged to an American astronaut. The agency rapidly recruited test pilots to take on the first flights, and on April 9, 1959, NASA announced it had chosen seven seasoned flyers from the ranks of the Marines, the Air Force and the Navy. The seven astronauts were introduced to the world just as Aikenhead wrapped up his affairs in Canada.

"We had to set an arrival date and sell the house in Brampton, along with the hundreds and hundreds of houses that were put on the market. But we were lucky, and we were able to sell. So we began to make the great trek, and we had to change planes several times," Aikenhead recalled, with he and his wife wrestling four children and a dog (who didn't want to get into the crate). They left their house at nine in the morning and didn't get to their hotel in Virginia until well after dark.[37] His new office at the Space Task Group at Langley was right next to the astronaut office. Talk about a change in fortunes.

It was hard to see our way clear to moon missions way back in 1959, but Aikenhead and his Canadian compatriots were there during the program's origin story. There's much more to say about their adventures, but it's already been covered in Gainor's excellent book *Arrows to the Moon*. Suffice it to say that Canadians (who, in some cases, were originally Britons) participated in practically every part of NASA's Apollo moon operations. Names like John Hodge (flight director), Tecwyn Roberts (chief of the Manned Flight Support Division) and Owen Maynard (chief of the lunar module engineering office) pepper Gainor's book — and other Avro veterans even went on to work on space station development.

So when you see Canadian astronauts decades later rising to

senior levels in NASA's human space program, do understand — they're far from the first to get there. Avro helped NASA get used to the idea of working together with foreign nationals. It introduced the agency to Canadian engineering excellence and our penchant for hard work. And Canada's work on Mercury helped us get all the way to the International Space Station 40 years later, although nobody could have predicted it at the time.

"The Avro group proved that Canadian and British engineers could work with the best America had to offer," Gainor wrote, adding that without the 32 Avro employees who arrived at NASA in 1959, "the road to Apollo 11's Tranquility Base landing [in 1969] may have been different and more bumpy."[38] Unfortunately, the Canadian space program they left behind had quite a few bumps of its own in the 1960s.

Arrow and Canadarm showed how hard it was for Canada to put together any space policy in the 60s and 70s, because no single department was responsible for space. In fact, it's still that way today — but what made it even more difficult decades ago was that we had no single space agency for civilian activities. And as the excitement of the first moon landing faded in the 1970s, the United States — one of our proposed major partners — began to focus on other priorities.

Initially, the United States and the Soviet Union were bound by a space race that kicked off when the Soviets launched the first satellite, Sputnik, into space in 1957. A race for major milestones quickly followed, with each country striving to prove its technological superiority to attract other nations' support. At first, the Soviets appeared to be ahead, launching the first dog, first man, first woman, first spacewalk and first two-person spacecraft, among other milestones.

But in the mid-1960s, the United States dedicated several years to the Gemini program, a series of missions that would practise the important activities of moon exploration in the relative safety of Earth orbit, such as docking two spacecraft. Spacewalks in particular were a sticky matter; astronauts struggled to accomplish even simple tasks such as moving from place to place on the hull of a spacecraft. More handholds, they kept saying when they came back to Earth. It took NASA several attempts and a few overheated astronauts before they created enough supports for Aldrin on Gemini 12.

This underrated but important mission happened in November 1966, which space observers say is about when the United States began to take the lead in the race, before Apollo 11 landed on the moon in July 1969. Why? The reasons are complex and have garnered entire books on their own, but in a recent magazine article, Soviet space expert Asif Siddiqi boiled it down to a few key factors.[39]

The first was engineering difficulties: the Soviets didn't even have a prototype of their mighty moon rocket ready until 1964, three years after President John F. Kennedy declared that the United States should "win the battle that is now going on around the world between freedom and tyranny." Referring specifically to the flights of Yuri Gagarin (first man in space) and Sputnik as examples, he called on US representatives, in a joint session of Congress, to achieve the goal "before this decade is out, of landing a man on the moon and returning him safely to the Earth."[40]

The Americans built their mighty Saturn V in the public eye, while the Soviets responded with their secretive N-1 rocket, which, at 105 metres, was almost as tall as the American Saturn V. Soviet chief designer Sergei Korolev was the lead engineer for the country's program and shepherded the rocket's development. Initially, Siddiqi pointed out, the N-1 was a big rocket with no firm mission — it could be used for military purposes or to launch a space station. It wasn't until July 1963 that Korolev focused on a lunar landing, and it took until August 1964 to convince Soviet leader Khrushchev to approve the idea.

Problem two was how the Soviet space program of the day was designed, Siddiqi said. "While NASA was a centralized, top-down system run by the federal government," he wrote, "the Soviet space program acted more like a socialist version of a competitive market. But rules were followed only half the time, and the program was held hostage by bureaucratic gridlock and the whims of powerful individuals."[41]

In particular, a rivalry arose between Korolev and Valentin Glushko, a rocket engine designer, about which rockets were most suitable for Soviet missions. Glushko focused on the terrifying ICBMs and thus wanted more storable propellant — a reasonable decision, given these missiles can sit around for months or years before any use. Korolev, however, wanted liquid fuel, a type of rocket propellant that can lift more mass.

While a Russian commission approved Korolev's plan over Glushko's in 1962, Glushko (with the help of influential friends in the Communist party) continued his work and even created a competing rocket project called the UR-700 in 1967. Since spaceflight is expensive, this bitter rivalry ended up not only defocusing the Soviets but costing a lot of money. This in part led to the third key trouble plaguing the Soviets, Siddiqi said, which was financial issues.

With Soviet departments competing against each other and the N-1 going ahead without full political support, this rocket underwent four launch attempts before even being tested on the ground.[42] That's akin to skipping all rehearsals before asking the National Arts Centre Orchestra to do a live performance in front of roughly 4,000 people — although engineering decisions like this have greater consequences than playing out of tune.

This hasty decision cost the Soviets the moon. On July 3, 1969 — only two weeks before the Apollo 11 crew made its historic launch towards the moon — the N-1 made its second and most infamous launch attempt. It rose to 100 metres above the ground, then began tilting strangely. Moments later, the behemoth collapsed on the Baikonur launch pad and exploded. While no fatalities were recorded, it was clear the Soviets weren't yet ready to reach the lunar surface with people.[43] They did try using a robotic craft, Luna 15, to pick up a sample of the moon's regolith (soil) just as the Apollo 11 crew walked on the surface. The goal was to return it to Earth before the astronauts got home. But Luna crashed, and the dream of Soviets becoming first on the moon died with that.

The success of Apollo 11 is usually regarded as an American achievement, but Canadians played a significant role — from the early conception of the STEM antenna to the Héroux engineers who created the lunar legs that let the Eagle lander safely alight on the lunar surface. Canada did justifiably celebrate its contributions to this space program, despite the political chaos plaguing its own efforts.

The first issue was, just as for the Soviets, a matter of limited funding. Canada is a large country with a small population, which always leads to tough decisions being made. Throughout history, as you read space plan after space plan, you see Canada choosing to pursue space in partnership with others. Our first satellite, Alouette, was a Canadian satellite launched on an American rocket, from an American launch pad. Our first astronaut, Marc Garneau, joined an American crew on an American space shuttle and also launched from an American launch pad. We find innovative ways to get our people and hardware into space, but it usually requires having somebody else fund the expensive launch equipment.

The second problem was lack of clear direction. Shortly after Alouette launched, Diefenbaker's cabinet (the same one that had cancelled the Arrow) remained unmoved when representatives from the Defence Research Board, the National Research Council and the Royal Canadian Air Force all lobbied for an official Canadian space agency. Frustrated Canadians with expertise in this field were moving south, and not just because the Arrow was gone. "The slower pace of Canadian aerospace technology development ultimately resulted in some of Canada's best and brightest moving south of the border to work in the much larger American space program," wrote Andrew Godefroy in *The Canadian Space Program*.[44]

One of many calls for a civilian space agency happened in the 1967 Chapman Report, led by the Canadian aerospace giant John Chapman who was so instrumental to space history here that our Canadian Space Agency headquarters are named after him. As a scientist and key administrator with what was then called the federal Defence Research Telecommunications Establishment, Chapman and his group did a survey of the various space research programs in Canada at the time, ranging from government to university to private industry. They called again for a centralized space agency, and a telecommunications satellite besides to help citizens of our huge country keep in touch with each other.[45] The effort was in vain, however: "Seemingly endless government departmental reorganization remained the bane of Canada's space program throughout the 1960s," Godefroy lamented.[46]

And the third problem, as in the case of the Soviets, was timing. By the time the report was published, NASA was already thinking past the moon landings, and the mood in Washington was different. The Soviet "threat" was already diminishing in the minds of American policymakers; by 1972, US president Richard Nixon and Soviet Premier Alexei Kosygin signed an agreement that forged a joint space mission later called the Apollo-Soyuz Test Project, which flew to great success in 1975.

On top of everything else, there were calls to make space a more profitable enterprise. In some ways, that thinking was four decades too early. SpaceX founder Elon Musk — he of the self-landing rockets to save on launch costs — was only born in 1971. CubeSats — those small satellites that can do Earth observation or science experiments at low cost — only became possible when computer miniaturization accelerated in the late 1990s. So, in the parlance of the 1970s, space profitability had to come from the government being responsive to customer needs.

That led to the space shuttle — a marvellous invention for the 1970s. Unlike the throwaway Apollo spacecraft, these space-crafts would launch into space, come back and then launch once again. The secret was using a new type of heat shield to protect against the forces of Earth's atmosphere during re-entry: a set of thousands of individualized tiles spread across the belly of the shuttle.

NASA touted the space shuttle as a way of saving money. One 1980 book cheerfully shows a picture of a satellite, proclaiming that "gathering data by satellite is speedy" compared with aerial surveys. The book says that the shuttle "will place satellites in orbit for one- to two-thirds the cost of launches aboard the Delta, Atlas-Centaur, and Titan rockets used for most recent US civilian and military missions." But that depended on the shuttle launching more than 50 times a year — about once per week.[47]

Spoiler alert: flights never became anywhere near that frequent. By the end of the program, the space shuttle launched about three to four times a year, and the promised cost savings from spaceflight were never realized. Other design factors made cutting costs even more difficult — a large payload bay to hold US military satellites and insistence on only partial reusability for the booster systems, for example. From a financial perspective, the shuttle never did reach its goal.

What it did achieve was a new rhythm to space explora-tion. Spaceflight became more routine, in the sense that crews went up and crews went down, and after a while, the media and public stopped paying attention to them unless something unusual or dangerous happened, which left space advocates unhappy. But with shuttles circling Earth on a regular basis, performing various sorts of scientific experiments with only minor variations, some journalists felt there really was not that much new to report. However fascinating it might have been for scientists and space journalists, the challenge remained: making it all relevant to taxpayers.

The shuttle's greater contributions, however, came in two ways. The first was allowing foreign nationals to fly on American spacecraft, which was crucial to the very existence of Canada's human space program. At first, our astronauts were limited to a small set of experiments and the basics of training, under a designation known as "payload specialists." As we gained experience and the growing trust of American partners, our astronauts were sent for more complicated "mission specialist" training. Spacewalks and robotics and a wider set of responsibilities came with the assignment, not to mention more prestigious ground positions. All of it helped Canada's space program mature.

The shuttle's second contribution was it enabled collaboration on the International Space Station. This monumental construction effort — touted by *Guinness World Records* as the most expensive man made object in history — emerged after a predecessor project never got off the ground. This earlier project, called "Space Station Freedom," represented a coalition of nations — Japan, Europe, Canada, the United States — who were dedicated to building a democratic, market-driven facility.

That all changed when the Soviet Union collapsed. Space shuttles began running missions up to former Soviet space station Mir, giving several Americans valuable experience in long-term spaceflight. Then the new Russian Federation joined the space station coalition, which would lead to the creation of the International Space Station. Shuttle flight after shuttle flight brought up the pieces, and Americans and Russians (as well as other nations) worked together "outside" to build the facility, spacewalk after spacewalk.

It was Canada's and the coalition's acceptance of the Russians that built the space station — and, to a large extent, built Canada's human space program. Only because our people accepted learning Russian, and living and training in their formerly secretive facilities, did Canada's astronaut program evolve from a handful of talented payload specialists to space-walkers and even a space station commander.

While Horner's unabashed jostling to meet the astronauts in 1969 may seem out of character for a Canadian politician, we should remember that when space is hot, everybody wants to get on board. Decades later, Prime Minister Justin Trudeau appeared at the Canadian Space Agency during Saint-Jacques's 2019 mission to make a special announcement: Canadarm had performed so well over the decades that Canada now plans to fund a Canadarm3. The reason? NASA wants to return to the moon, it has presidential support to do so, and Canada doesn't want to lose 50-plus years of industrial expertise in building the epic Canadarm line. Trudeau's appearance did get media attention, although, as we'll explore later, his timing was curious for that had nothing to do with space.

CHAPTER 2

"Three strong Canadian arms on board"

> Where is the man who can clamber to heaven?
> — *The Epic of Gilgamesh* (translated by N.K. Sandars)

Marc Garneau found a spot on the deck of a 59-foot sailing boat, which was no small feat with a dozen other people on board. He and his crew were in a transatlantic race from Newport, Rhode Island, to Cork, Ireland, and had a rare moment to relax while floating in the calm English Channel.

A waxing crescent moon illuminated the sky, and the crew kept looking at it while listening to broadcasts on a short-wave radio. It was there that Garneau heard Neil Armstrong's historic pronouncement as Armstrong stepped on the moon: "That's one small step for man, one giant leap for mankind."

The 20-year-old Garneau — then a student at the Royal Military College in Kingston — was just beginning his career. "I can tell you, we were just flabbergasted, all of us," he said.[1] "I thought, 'Wow, it's amazing. That you look at that moon up there, and somebody's actually getting out of the Eagle to walk on that moon.'"

And as the man who would become the first Canadian in space listened to America's first man on the moon, one would think that maybe there was a spark — a realization that this would be an interesting path to take. But that's a Hollywood wish, not real life. "Of course, I had no idea that one day I would get to go into space," Garneau added. "I watched it like everybody else, and I listened. It was just a radio for us."

Yet there is an interesting historical footnote to Garneau's sea journey. When it came time for mission managers to make a choice for who was first to fly in Canada's space program, he argued it was probably this voyage — which he actually performed twice — that helped their choice.

You see, being on a small sailboat — or even a larger Navy vessel, which Garneau was many times in the early years of his career — has parallels to spaceflight. Sure, you can breathe the fresh air whenever you go on deck. And it's not as though you need to ride a rocket to get to your destination. But in many ways, the crew dynamics are similar. You're isolated, you're dependent on your shipmates and you work in dangerous conditions.

"You have to fix things if you can, if it's possible, if they don't work. So I thought that I had the kind of temperament for it," he recalled. It was only 15 years later that Garneau took another epic voyage, this time aboard the space shuttle Challenger — the only Canadian in an otherwise all-American crew. He was the first Canadian man in space, on the first crew to host two women and with a crewmate that was the first Australian-born person in space. Garneau, just 35 years old, became famous in Canada instantly.

Garneau accepts his mantle as Canada's "First Man" — to borrow from the Hollywood biopic about Armstrong that played in 2018 — and has learned from the astronaut experience, he told me. He found his time in politics, like space, has included quick-moving situations that require measured responses.

When I interviewed Garneau in his beige-walled Montreal constituency office, he was Canada's minister of transport and in the thick of the 737 MAX 8 crisis. In the wake of an Ethiopian crash, Canada had joined the rest of the world in grounding its domestic fleet of what were supposedly cutting-edge airliners until the cause of the accident could be identified. I journeyed from Ottawa for my interview in Montreal only two days after the planes were grounded.

It was necessary standard procedure, as aviation investigations go. It protects public safety. But it was massively inconvenient, nonetheless. Thousands of Canadian passengers were delayed or stranded and clogging airport help lines. Garneau himself made the announcement on national television and was, staff told me, still in meetings dealing with the situation. When my planned one-hour interview was cut to half an hour, I completely understood; I was very lucky to see him, under the circumstances.

At age 71, Garneau continues pulling long days as a member of Parliament and even shows up at space announcements from time to time. It's a tough job with high public accountability; in 2011, when the Liberals had their worst-ever showing in Parliament with 34 seats,[2] Garneau won his large suburban riding by only 642 votes.[3] Yet observers say Garneau deals with transportation and political crises methodically. When reporters ask him tough questions, little rattles him. Minutes into his interview with me, Garneau brought up his first wife, who had died by suicide 30 years ago. I gently asked him if his sudden fame might have contributed, and his confident answer was just as gentle: he was sure that wasn't a factor.

"I think I knew my wife well enough that I could separate what the cause of her death [was] from the change in our life," Garneau said, adding that his wife had been a steady force for him and their twin eight-year-olds through the tumult of training, the scrutiny during the launch and mission and the never-ending mob of reporters asking for his attention after he landed.

Hollywood's *First Man* depicted a troubled Armstrong (played by Ryan Gosling) who seemed only one crisis away from a breakdown. His three-year-old daughter died of cancer. His marriage was on the rocks. Movie-Armstrong was therefore distant and could only deal with the pressures by going outside and staring at the stars with a telescope.

It's a compelling movie story, but the reality was more subtle. Armstrong and his then wife, Janet, insisted on keeping things as normal as possible for their children, before and after their divorce. Armstrong also did his best with the overwhelming fame (which, for perspective, included such weirdness as a barber selling Armstrong's hair clippings for $3,000 in 2004 — a quarter-century after Apollo 11's mission).[4] The media bemoaned Armstrong's tendency to decline interviews that didn't interest him, forgetting that Armstrong still showed up in public; he served on space committees and starred in Apollo 11 anniversary celebrations decades after retiring from NASA.

"My parents both, I think, tried very hard for things not to change," Armstrong's son, Mark, told me for a Space.com interview in 2019, shortly before the 50th anniversary of the historic flight. Mark was only six years old when his dad bounced around the Sea of Tranquility with Aldrin. Armstrong died at age 82 in 2012; today, Mark gladly gives interviews to talk about his father's legacy.

"They wanted our family dynamic and the people that we were," he added, "to be the same after the flight as before the flight. I think Dad had the worst challenge [of the astronaut corps] — the most difficult challenge, probably by several orders of magnitude — because of the number of requests and requirements and opportunities and mandates that were coming at him on a daily basis, which were overwhelming."

The "First Man" mantle follows Garneau decades later, too — even though he made his last flight in 2000, nearly 20 years ago. He has since served as president of the Canadian Space Agency, chancellor of Carleton University, and a Liberal Party member of Parliament. I was curious whether, after all this time in other public positions, his constituents still recognize him as an astronaut first. They mostly do, he said, even after more than 10 years serving the same riding and five years as transport minister. He still can't go to a restaurant without getting attention, but he must be used to it. It's been almost 40 years since he had a quiet meal.

"Our astronaut program didn't just materialize out of thin air, although ironically you could say it was beamed down from the Enterprise," Wally Cherwinski told me.[5]

Cherwinski was doing public affairs for the National Research Council, a position he arrived at through a genuine interest in space; while Garneau was floating on the English Channel listening to Armstrong, Cherwinski and his buddies at the University of Western Ontario (now Western University) watched on a black-and-white television. It was only years later that he learned that future astronaut Roberta Bondar was doing the same thing just across the street from him.

In spring 1983, the Enterprise landed in Ottawa. Not the starship of *Star Trek* fame but the space shuttle named after it. Enterprise wasn't designed for space. It was a space shuttle glider for astronauts to see how the vehicle would behave during landing. NASA completed its glider testing in the late 1970s and then sent Enterprise on goodwill tours, riding piggyback on a Boeing 747.

"I remember standing on the roof of a building at the Ottawa airport, and I was doing a play-by-play on live radio with the local radio stations as it circled and it landed," Cherwinski said.

"I could see cars stopped on the road into the airport, and there were already thousands of people in place, and they were cheering as it came in. It was here on the ground for two or three days at the airport, and it drew enormous crowds. Traffic jams every day, and we guessed that about three-quarters of a million people came to see it. Now, you couldn't actually get up close and touch it, but just getting near enough to get a good look was important. It was important enough to people for them to come."

While the crowds gaped at Enterprise, NASA administrator James (Jim) Beggs made a special announcement in Ottawa: the agency would ask Canada to send Canadian experiments into space, along with a Canadian astronaut, as a thank you for the success of Canadarm. By the time Enterprise beamed into town, the shuttle arm had flown multiple times and proven its worth many times over, between releasing and retrieving satellites and once even knocking ice off the payload bay. "Clearly we had impressed them with Canadarm, and they knew that they would have a skilled and reliable partner in other space activities," Cherwinski said.

This actually wasn't the first time that NASA tried to get Canada interested. The original 1974 memorandum of understanding about Canadarm between NASA and the NRC promised Canadians access to the shuttle. The agreement, however, only talks specifically about matters such as space for experiments — not space for astronauts.[6] But after the agreement was signed, NASA created a special kind of astronaut designation called a "payload specialist" that would be fully responsible for a company's or foreign country's experiments. Countries contributing technology to the space shuttle program (or to whom NASA wanted to extend goodwill), or buying a launch for their satellite or experiment, would thus have a chance to fly their citizens into space.

Canada got its astronaut offer in August 1979, before Canadarm actually flew. NASA was so impressed with early tests of the arm that it wrote Chapman (of the Chapman Report, who was then former chair of the federal committee on space) offering a seat for a Canadian aboard the space shuttle. But the invitation appeared to get lost, as Chapman, just 58, died that September.[7]

"We have no record of ever receiving a reply," NASA spokesperson Lyn Wigbells told the *Ottawa Citizen* in September 1981. A space official at the NRC, Art Hunter, told the *Citizen* he wasn't even aware that the invitation had been issued.[8]

If the invite had been accepted, Canada could have been involved in the very first training program in 1980, John Sakss (of NASA's international office) told the *Citizen*, digging the metaphorical knife a little further under Canadian skin. But another training program was set to begin in 1983, Sakss said.

In typical careful public-relations fashion, Sakss did not mention (at least, as far as the article tells us) whether NASA was prepared to reissue the invitation to Canada. But the chagrined NRC seemed very interested, as far as a facts-only journalistic article can say. "NRC's Hunter says he will pressure the government to accept NASA's offer," the *Citizen* added laconically.

The buzz about Canadians in space persisted that year. At the end of 1981, NASA astronauts Engle and Truly were in Montreal as part of their six-day Canadian tour, celebrating the first flight of Canadarm. Astronauts are also carefully trained in what to say before they are set loose on the public, yet even these men told the crowd that "there is no reason Quebecers can't become astronauts in future manned space programs," in the words of the Canadian Press.[9] ("First Man" Garneau, coincidentally, was born in Quebec.)

It's unclear how much NRC pressure actually took place in 1981, but nevertheless, Canada was given its chance to bring astronauts to NASA in 1983. So how did they recruit said astronauts? In typical Canadian fashion, it was understated:

a simple "Help Wanted" ad in several Canadian newspapers, running casually beside more Earthly job positions.

"Of course, the media picked up on it and they ran stories about our ad and the astronaut search, and it just took a life of its own," Cherwinski said. "We set up a selection committee and went across the country to interview candidates. Naturally, wherever we went there would be regional media who would provide a lot of coverage, especially when there were local applicants involved."

But some people still heard about the ad through happenstance. Robert (Bob) Thirsk was working as a medical doctor in a small town in New Brunswick. One Saturday morning, he finished the rounds and wandered into the doctor's lounge. There, he began browsing the newspaper (as one did in the 1980s) and came across the ad.

"This Grade 3 dream came flooding into the front of my brain," Thirsk said,[10] adding that he never just wanted a career — he wanted to make a difference and to have influence. While being a doctor checked most of the boxes, being an astronaut was something more.

"My thinking is that if you're going to have a satisfying career, you need to do something that you're good at. Something that you're passionate about, you believe in," he said. "And also something that makes the world a better place. I regarded a career as an astronaut as something that met all those criteria."

So Thirsk applied, and Garneau, and 4,000 other Canadians. It took several months to whittle down the pool to six candidates who would go for final astronaut training. The final group included neurologist Bondar, Canadian Navy commander Garneau, laser physicist Steve MacLean, physiologist Ken Money, medical doctor Thirsk and NRC aerodynamics research officer Bjarni Tryggvason.

Canada never had the same veneration for its first astronauts that the Americans did. The Mercury 7 were immortalized in breathless, 1960s-era exclusive profiles in *Life* magazine, and this attention persisted for newer generations of astronauts recruited for Gemini and Apollo. Even today, the Apollo moon astronauts bring in crowds for autographs more than 50 years after the first moon landing. Nevertheless, this list of Canadian names — dare we call them legends? — evokes stories for space buffs in our country.

One extraordinary thing about Canada's astronaut corps is the degree to which many of them remain in public service many years after leaving space behind. Garneau is just one example. Today, Bondar runs a wildlife foundation, and Thirsk is constantly available to students and the public at events throughout Canada. And Money (who never flew in space), MacLean and Tryggvason heavily pursued research efforts after leaving the Canadian astronaut program. Like sharks, these people — already famous as researchers and astronauts and all in their 60s and 70s today — cannot stay still.

Garneau was selected for flight in March 1984, just months after he and the other Canadian astronauts were brought into the program. No one expected the opportunity would come so quickly. Training for Canadians was very different then. Most of the training took place in Canada. Even after Garneau was assigned to his flight, he and his backup (Thirsk) only had a limited time in Houston to get to know the crew and his assignments. Today, by contrast, Canadians train side by side with their American (and European, Russian, Japanese and so forth) counterparts for many years.

At first, the Canadians had a lot of convincing to do, recalled Charlie Bolden, an astronaut at the time of Garneau's selection. "For some reason, the astronaut office was very insular. I think we felt much more highly of ourselves than was deserved," he said. "We felt — I'll use the term 'we' [even though] I was not one of them, but I was in the office — we felt that we were so special that we didn't want these outsiders, these payload specialists, to come in and rain on our party. So, that was sort of the atmosphere when the first group of Canadians came down. I think we got over it in time. Thank goodness."[11]

A recruit today isn't even called an astronaut for a good two-and-a-half years, at least not until they pass basic training. Due to the amount of training needed for six-month International Space Station flights, many American astronauts must wait at least five or six years before flying into space.

A one-week space shuttle flight in the 1980s or 1990s or 2000s, however, usually didn't need nearly as much lead time. (Williams, while reading over an early draft for this book, pointed out that his STS-90 mission was so complex that his crew required two years — but that was unusual.) Astronauts flew sooner and more frequently, so much so that in the early 1980s astronauts were often shuffled between shuttles. This happened to Garneau; by early June, amid a larger NASA change of several crews and experiments, Garneau was assigned to STS-41G and the space shuttle Challenger, instead of the original flight on Discovery. He also was slated to fly three weeks earlier, adding a little more pressure to his training schedule.[12]

Garneau pointed out that as payload specialist, however, he didn't need a high degree of training for all shuttle systems, although certainly training was necessary. "Obviously, I had to receive enough training not to be a liability to the others; to know what to do in case something went wrong," he said. "How to use the communication equipment, how to go to the bathroom, how to prepare meals. All of those things that everybody has to do. Most of my focus was on my experiments and how I would do those experiments during the time that I was up there."

Three weeks before the flight, Garneau and his crewmates were invited to the White House to meet with President

Ronald Reagan (who was pushing for the as-yet-unnamed space station at the time) and newly elected Canadian prime minister Brian Mulroney. One account says that Mulroney was trying to get a meeting with Reagan quickly and used an astronaut to get a foot in the door,[13] although Garneau describes the situation differently.

Rather, Garneau recalled that the high-level conversation between the US and Canada was to show good relations between the countries. The response from Canadian officials: "Oh, there's a Canadian training at the moment."

For Garneau, the pressure didn't come from this sudden diplomatic visit, or even from performing on the shuttle; from his training, he was ready for these situations. What worried him more, he said, was: "Am I going to be able to live up to the high expectation the Canadians have?"

Reagan was also new to spaceflight in 1984, although his interest came from the diplomatic angle — a desire to show the economic superiority of the US and its allies in the face of what he saw as a larger threat. One of the first hints of this came in his March 30, 1981, speech to the National Conference of the Building and Construction Trades Department — a speech focused on labour and the economy, to be sure.

Reagan was just two months into his first presidential term and still looking for a way to define himself. And near his speech's end, he shifted tracks — he suddenly began talking, to this group of people responsible for America's infrastructure, about his worries concerning the Soviet Union.

"Since 1970, the Soviet Union has undergone a massive military buildup, far outstripping any need for defence," he said.[14] "They've spent $300 billion more than we have for military forces, resulting in a significant numerical advantage in strategic nuclear delivery systems, tactical aircraft, submarines, artillery and anti-aircraft defence. And to allow this defence or this imbalance to continue is a threat to our national security. It's my duty as president, and all of our responsibility as citizens, to keep this country strong enough to remain free."

The speech would be Reagan's last public activity for a while. Reagan stepped outside of the building with his usual security entourage, and as he passed a stop sign, he raised his hand and smiled at the crowd. Then came the gunshots. Personnel pushed Reagan to the sidewalk,[15] but he and three others suffered gunshot injuries of varying degrees of severity. Reagan recovered relatively quickly from a lung puncture and internal bleeding. The shooter was later apprehended, arrested, charged and put in jail.

Reagan continued to perform many of his presidential duties while recovering, famously among them corresponding with Soviet leader Leonid Brezhnev. A sensitive letter from Brezhnev reached the senior level of US government on March 6 and was translated overnight.

"The Soviet Union has not sought, and does not seek, military superiority. But neither will we permit such superiority to be established over us," Brezhnev wrote in the now-declassified letter.[16] "Such attempts, as well as attempts to talk to us from a position of strength, are absolutely futile." He urged Reagan not to seek an arms race or an atomic war in dealing with the Soviets, calling these tactics "dangerous madness." He added that the Americans and Soviets were in "imperative need for the conduct and development of dialogue that is active and at all levels." While Reagan was urged not to make this letter public, more and more the Soviets became a theme in the president's speeches as his first year progressed.

Forty-four days after Brezhnev's letter, as Reagan convalesced, the US president wrote his response. "Wrote a draft of a letter to Brezhnev. Don't know whether I'll send it [sic] but enjoyed putting some thoughts down on paper," Reagan noted in his diary. Reagan's letter pointed out the "USSR's unremitting and comprehensive military buildup" as well as the country's "pursuit of unilateral advantage in various parts of the globe," themes he would continue to discuss publicly during his presidency. He also expressed the need for relations between the two countries to improve, but called on the Soviets to exercise good intentions through "the basis of actions and demonstrated restraint." Still, he added, "We should work together to avoid misunderstanding or miscalculation."[17]

Martin Anderson, a senior economic advisor to Reagan, later co-authored a book in which he pointed out that Reagan's tone was "significantly friendlier" than the one used in a draft the US State Department proposed, although Reagan did say Brezhnev was allowing government ideology to overrule the people's right to the "dignity of having some control over their individual destiny." This gentle poke at the Soviets would get stronger over time, Anderson said, particularly when Reagan began equating the Soviets with modern evil.[18] Nevertheless, the unique approach Reagan took to speaking with the Soviets, Anderson argued, was likely the "beginning of the end of the Cold War."[19]

One of Reagan's first moves — in secret, and then in public — was to pledge to stop deploying Pershing II and ground-launched cruise missiles, but only after the Soviets agreed to remove some of their own. "This, like the first footstep on the moon, would be a giant step for mankind," he told the National Press Club in Washington, DC, on Nov. 18, 1981.[20] But another matter was weighing on his mind — the possibility of the Soviets launching attacks from space, rather than the ground.

"Had a briefing on the Soviets & Space [sic]. There is no question but that they are working (twice as hard as us) to come up with a military superiority in outer space," Reagan wrote in his diary on Aug. 8, 1983.[21] That December, he made another diary note related to Soviet space policy, but this one focused instead on his response to NASA's budget. "I think we're OK there & can still start to plan a space station."[22]

Reagan was active in civilian space even before Space Station Freedom was announced. He asked his aides to get him up early on April 12, 1981, the day after his return from the hospital. Reagan was so ill from the shooting two weeks before that he could only work for an hour at a time, but he still wanted to see the launch of the very first space shuttle — Columbia — for a successful two-day shakeout mission in space.[23] He was entranced with the two astronauts — moon veteran John Young and rookie Robert Crippen — and presented them with the NASA Distinguished Service Medal in March, before they took flight.[24]

Yet Freedom wasn't in Reagan's mind in 1981 — it would take others to put it there. And the work had been going on long before he came to the Oval Office. A permanent space station, in fact, had been in NASA discussions for more than a decade. In February 1969, as preparations for Apollo 11's moonshot ramped up, President Richard Nixon appointed an ad hoc Space Task Group to decide what to do after the Apollo program finished.

NASA, used to open purse strings at the height of the Cold War, now faced a different environment as the Vietnam War dragged on. One estimate pegs the Vietnam cost at $2 billion a month in late-1960s dollars; the budget deficit in 1968 was $25 billion.[25] NASA's plans for successively more ambitious Apollo missions — under a program known as the Apollo Applications Program — never came to fruition.

The genesis of Freedom goes back to this era, when NASA was trying to figure out what to do next. The Space Task Group (led by the US vice-president) created an ambitious 1969 plan that, among other items, included a massive 12-person space station. (Today's International Space Station routinely hosts six crew members; while 13 have been on board during two separate missions, it was only for a few days and only with an attached space shuttle ready to provide extra capacity.)

Unlike with Apollo, however, NASA was determined not to take on this new plan alone. "The principle that there would be international participation in whatever might follow Apollo came first," said John Logsdon, the founder and former director of the Space Policy Institute at George Washington University,[26] now semi-retired at 82 years old. (His legacy includes serving on the investigatory board after the Columbia space shuttle broke up during what was supposed to be a routine re-entry in 2003. His board found that NASA ignored the danger of tank foam that broke a hole in the shuttle's wing during takeoff, creating a breach big enough for hot plasma to enter and melt the wing's internal structure during landing.)

In the restricted-budget era of the day, however, NASA made a mistake in trying to ask for so much money at once — its plan included a space station, a fully reusable space vehicle and plans for sending astronauts to Mars in the 1980s. Funding scaled back the space shuttle to a partly reusable vehicle instead. Since the account of Space Station Freedom and its failure has been covered in many other excellent volumes, we will not get into the details here.

Garneau was far too busy to focus on distant space programs in 1984. His focus and dedication served him well. On Oct. 5, 1984, the light emanating from space shuttle Challenger's launch split the pre-dawn darkness at the Kennedy Space Center in Florida. Mission 41G was on its way to space for eight days. And Garneau was catapulted into even more fame, fame that would put him into the history books as Canada's first person in space.

Reporters of the day were enthralled by the flight and wanted Garneau to call out vivid descriptions of his spaceflight to the ground. When he said little, they started to call him "The Right Stiff" — a pun on the title of Tom Wolfe's rollicking book describing the larger-than-life Mercury astronauts of two decades earlier, who were chosen because they had "the Right Stuff."

The movie version of Wolfe's account had hit theatres just a year earlier, and Garneau points out that standards of space celebrity were different then. There was no Twitter and no David Bowie music videos from space, as Hadfield would use decades later. Forget about the internet, which existed, although most Canadians had never heard of it and didn't have access to the first networks. Heck, there was no continuous communication with the space shuttle. Challenger's orbit bounced the signal from satellite to satellite on NASA's dedicated Tracking and Data Relay Satellite system.

It is also important to remember the pressures of a two-week space shuttle flight, compared to today's typical leisurely six-month lope aboard the International Space Station. There simply wasn't time to relax, to emote, to send back information to Earth. Every moment counted, and when you were the junior member of the crew, you followed your commander's orders. Garneau pointed out that he was not tasked with much radio communication anyway — why clog the channel with useless chatter?

Garneau was indeed very busy: responsible for 10 Canadian experiments that lay the foundation for future space technology, space sciences and life sciences.[27] One of the most famous was tests of the Space Vision System (SVS), a computer vision framework that would, years later, be adopted by Ottawa-based space engineering firm Neptec and — many years later — would help build the International Space Station. Lighting is weird in space, with harsh shadows and no atmosphere, making it hard to judge distances, as astronauts discovered when doing their moonwalks in the 1960s and 1970s. SVS allowed for millimetre-scale precision when putting pieces of the space station together. And that flows back to the first experiments performed by Garneau.

Some of Garneau's other experiments included how the human body adapts to space (an ongoing obsession of doctors since the first person flew in 1961) and examining physical characteristics of space and the Earth's upper atmosphere. So he had much to do. Yet the Canadian press, largely unaware of the rigours of astronaut-hood, wanted their man to emote about the scene before him, even asking for descriptions at an in-flight press conference of how he felt and what his impressions were of spaceflight.[28]

Garneau got a chance to speak with Reagan once again, when the president made a four-minute phone call to the crew on Oct. 12 — which Reagan mostly used as an opportunity to talk about the Wright brothers, since he was standing before a crowd in Dayton, Ohio, where the first US planes were built in the early 1900s. Reagan talked individually with several crew members, including Garneau.

"A lot's happened since we talked last at the White House," Reagan said to Garneau, then added a quick reference to the Canadarm: "With all there is to do in this mission, I know that Cripp [commander Crippen] appreciates having three strong Canadian arms on board."[29]

Garneau completed his work, and as the expression always goes, he arrived safely back on Earth. Upon returning to his training centre in Houston, Garneau was temporarily swarmed by an excited surge of Canadian journalists. He calmly answered questions, backed against a plane with his young twins clinging to his legs in confusion. Ten minutes later, the crowd dispersed as reporters stampeded to pay phones to file their stories.[30] (Again, remember, there was no internet as we know it today.)

But in a pattern typical of Canadian astronaut stories, the attention on Garneau eventually faded. Another Canadian was picked for a flight, MacLean, and he also buried himself in training. Space quickly fell off the headlines, and Canada became interested in other matters. It's a pattern that continues today, which is difficult for the astronaut corps since many years now pass between flights. It might be months or years before their names are even mentioned in major media these days.

Why ignore the program in between launches? It's easy for reporters and the public to follow astronauts during the fire and fury of a launch, and to watch them float in space for a few days or weeks or (in later decades) months. What's harder is to watch their evolution over decades. To see the incremental work that they put in to make spaceflight better. This time in training and mission support is most of an astronaut's *life*. Yet the work they do here is mostly unacknowledged, unchronicled and unremembered.

Some people think that astronauts spend their time in between flights waiting around for the next opportunity. Actually, the work never stops because a job of an astronaut is always to support his crewmates, whether on the ground or in space. So while the world looked elsewhere, Garneau leveraged his military simulation experience to build better caution-and-warning systems on the International Space Station. His work manipulating Canadarm in later missions literally helped build the station, too.

Garneau served as a CAPCOM on 17 space shuttle missions — he was "First Man" for Canada there too, in fact. CAPCOM is supposed to be helpful and calm even in the worst situations. Shortly after an explosion aboard the moon mission Apollo 13 in 1970, Jim Lovell called down from space: "Houston, we've had a problem." He and the CAPCOM astronaut, Jack Lousma, briefly discussed a warning that appeared in the spacecraft. Then Lousma laconically replied: "Okay, stand by, 13. We're looking at it."[31]

Crewmate Fred Haise quickly described problems such as "a pretty large bang," a "jolt" and an "amp spike" — and each time, Lousma's response was a calm "Roger" before relaying the first instructions to address the issue.[32] Mission control was a flurry of activity as the team worked to bring the astronauts home safely, but the voice of Lousma and other CAPCOM kept a confident tone to help the crew focus.

Mission Control succeeded, by the way. That's why the 1995 movie *Apollo 13* exists. It's because Lovell survived, thrived and (with journalist Jeffrey Kluger) wrote a book about the mission upon which the movie was based.

Apollo 13 may have made for good Hollywood fodder, since everyone came home safely despite the dangers. But that doesn't always happen in spaceflight. A 1986 flight of Challenger — Garneau's same spaceship — was a stark reminder of what can happen when complacency sets in.

CHAPTER 3

Space cards

The best way out is always through.

— Robert Frost, "A Servant to Servants" (1915)

It's called "normalization of deviance."

Spaceflight accidents, just like airplane accidents, are complex beasts. But one of the overriding themes seems to be accepting something abnormal as normal — even when it could be dangerous. The Apollo 1 crew in 1967 perished in a deadly fire on the launch pad inside of a spacecraft plagued by constant design changes. These were so frustrating that crew commander Gus Grissom had hung a lemon, perhaps plucked from his own backyard, on the command module simulator.[1] Grissom and his two crewmates died when a spark — perhaps from exposed wiring beneath the door — ignited and rapidly spread through the oxygen environment, accelerating through a spacecraft furnished with Velcro and other flammable items. NASA spent 18 months investigating and making design changes before the next crew flew.

And this happens again and again and again. In the American program, seven crew members died on the Challenger shuttle in 1986 when a cold snap stiffened a booster rocket seal, causing it to fail, allowing flames to damage the fuel tank during launch. Numerous engineers had raised concerns about the effect of the cold on the boosters, concerns that were put aside.

In 2003, the Columbia shuttle broke up during re-entry due to a launch incident nearly two weeks before — a piece of foam insulation from one of the external fuel tanks flew off during the launch and slammed into the vehicle's wing. The impact produced a roughly suitcase-sized hole in the leading edge of the port wing, letting in superheated gases and causing catastrophic failure of the wing and vehicle when the orbiter re-entered the Earth's atmosphere. Foam fails were a common thing in the shuttle program for years and were generally accepted as okay — though the external tank was supposed to be more secure. Afterwards, NASA made several design changes to the shuttle and implemented a new procedure to scan the entire shuttle to look for any damage in space, using a modified Canadarm with a Canadian-built camera by Neptec on the end.

To be sure, spaceflight is dangerous, and that cannot inherently be changed. But this is why astronauts spend countless hours simulating how to do things in a dangerous situation, instead of just focusing on how to do things when everything is going well. This is why many astronauts have backgrounds flying jets or working in remote areas or otherwise taking on careers with dangerous components, because you need that experience while flying in space. Just ask the two crew members on that Soyuz abort in 2018, which ended up finishing safely — it could have gone much differently in untrained hands.

But to dial the timeline back to the mid-1980s, there was a lot more optimism about the space shuttle's potential. NASA thought it could run dozens of flights a year and bring the cost of spaceflight down. The public thought that the space shuttle was safe, so safe that few were worried when NASA announced it would put a schoolteacher (Christa McAuliffe) aboard the fateful Challenger flight — this after flying a US senator, a US congressman and a Saudi prince as payload specialists. So it came as a huge shock to everyone when on Jan. 28, 1986 — 19 years, almost to the day, after Apollo 1 — Challenger exploded in front of millions of viewers, including some of McAuliffe's own students.

NASA carefully considers the lessons learned after each incident and works to make sure that spaceflight is safer for the next group. In mid-2019, I was working on a story for Space. com about spaceflight "contingencies," or those bad days in space that could lead to fatalities. There was some concern that NASA's recently announced accelerated push to land humans on the moon in five years could lead to rushing things and leaving aside safety. I called administrator Jim Bridenstine's office, and he promptly left me a voice mail explaining how the agency constantly checks its plans with internal and external experts to make sure safety is being considered. "Just know, in no way does NASA intend to mitigate safety at all when it comes to meeting the objective," he insisted.[2]

Each of these incidents — which I have only described briefly here — could generate books on their own investigating the causes and remedies. NASA has thick public documents available on each incident, too (with the exception of the Soyuz one I mentioned, which is a Russian-led investigation). The agency surely does its best to recover from incidents and to prevent them before they ever occur. But that did not lessen the shock in January 1986, when the Canadian astronaut corps only had one flight in the books and one more flight coming up. History would record a more than six-year gap before any Canadian flew in space again due to the investigation and a backlog of more urgent flights for US astronauts. Another two-year hiatus came after the Columbia accident in 2003.

Both of these times were crucial in Canadian spaceflight

history and policy. The period in between these two disasters saw the greatest run of Canadian space missions ever, with astronauts flying at least once a year for nearly a decade. But when the pace slowed again after Columbia, Canada's interest in spaceflight was on the wane for many reasons.

At the time that the Challenger accident occurred, Canada only had one astronaut assigned to a flight — MacLean, an Ottawa-born physicist. While there was no word on when his flight would actually go into space, MacLean, Tryggvason and Garneau all worked together to continue getting ready for flight. The primary goal of this flight was to test the Space Vision System to help astronauts see and snag objects more easily with the Canadarm; MacLean used the extra time to manage a program developing the next stage of SVS, which would use lasers and three-dimensional video imaging. At the time, they hoped it would be used on the space station — and indeed, a more advanced version of this system helped Canadarm2 construct the International Space Station in the 2000s.

MacLean told journalist Lydia Dotto at the time that the delay was somewhat beneficial for his flight, as he had six other experiments to worry about, which now had more time for development. For example, MacLean was testing a device to examine chemical processes that affected the ozone layer. This was a lengthy task, requiring him to board a commercial airliner and then try out the experiment as it flew over the North Pole. "Every experiment has improved tremendously because of knowledge we've gained in the interim," he said. "Many people don't appreciate what you're able to accomplish even though you don't fly in space."[3]

That remark is true, as many people forget that astronauts spend years in between flights not only supporting other missions, but also working to improve various facets of spaceflight. During a more than two-year hiatus after the Challenger accident, the work on the ground continued. Tryggvason waited the longest of the group, working for about 13 years after his selection on the ground before flying on space shuttle Discovery for STS-85 in 1997. So, in 1986, it would be another 11 years until he flew — not that anybody knew that at the time. In fact, there was a lot of uncertainty, he recalled, about whether the shuttle would fly again.

"As things transpired, the shuttle flew again in just under three years, but there was close to a five-year delay for the two Canadian flights," he said. "This opened time for the astronauts to work on projects of their choosing."[4] For Tryggvason, this allowed him time to deep-dive into the behaviour of fluids, which was puzzling in the 1980s and is still puzzling today. In fact, one of the difficulties in investigating the spacesuit leak of 2013 was that the dynamics of cooling fluid are not well understood in spacesuits, although of course NASA has made progress over the decades.

One can argue that Challenger forever altered Tryggvason's career, because it led to a long-standing fascination with fluid dynamics in space that defined much of his work in the following decades. He became so interested in it that after a single flight in space, he refocused his astronaut career on experimentation rather than flying — and then ended up leaving.

Tryggvason wanted to know which experiments were being planned to study fluid behaviour in a free-fall environment, he said. "The first part of that was to come to understand the disturbance that experiments would be subjected to by spacecraft vibrations and attitude control. Surprisingly, there was little attention on that, but the little information available did suggest that fluid experiments would be disturbed by spacecraft vibrations."

That intrigued Tryggvason, so his team began using the "vomit comet" — a version of the familiar Boeing 707 airliner that had been modified for military use, in this case making steep climbs and dives to simulate weightlessness. He observed that there was a lot of disturbance to the experiments, since they were only in free fall for about 20 seconds at a time. While this was a more extreme disturbance than what would be needed for space, "this led to the idea of developing isolation systems," Tryggvason said. And this engineering project would occupy much of his time on the ground for the coming decade.

The first of these projects was one he created with the University of British Columbia (UBC) for a single-degree-of-freedom motion isolation system, which flew on the vomit comet around 1990, he recalled. Tryggvason then had to put the work aside since MacLean was due for his shuttle flight in 1992 and Tryggvason was assigned as the backup astronaut should MacLean be unable to take on the flight.

Tryggvason had a heavy schedule during these years — he developed the experiment plan for the test of the vision system, he led the design and manufacture of a target assembly that could be maneuvered by the Canadarm for the vision system to lock on to, and he even helped develop full-scale models of the hardware used in the space shuttle simulators at NASA's Johnson Space Center in Houston, where astronauts train for missions.

"These roles were, for me, a crash self-educational course on designing space hardware, guided by the many NASA documents available," Tryggvason recalled. And his work continued even during the mission, when he was the principal interface between the CSA support team and the NASA mission control team.

But the isolation systems exerted a fascinating pull on Tryggvason, who dove back into work shortly after MacLean landed — with successful work accomplished on the Space Vision System — in fall 1992. Working again with UBC, Tryggvason developed a large motion isolation system and flew that several times on the vomit comet. This was a system that was completely insulated from vibrations, using magnetic levitation. NASA was so impressed that it agreed to fly the system on several space shuttle flights on the parabolic flight aircraft without charging any flight costs, which was a large win for Canada's small space program. NASA was also impressed by the mounting hardware that Tryggvason and his team designed for the target assembly, and offered to fly that on six space shuttle flights and provide opportunities to fly additional Canadian science experiments on the shuttle.

A more advanced version of Tryggvason's magnetic levitation device, known as the Microgravity Vibration Isolation Mount (MIM), also flew on the Russian Mir space station between April 1996 and January 1998, during an era when US space shuttles regularly ferried American astronauts to and from the Mir space station for long-duration stays. While many Canadians know that Chris Hadfield visited Mir in 1995, far fewer likely know that a Canadian experiment spent 2.5 years on that space station after Hadfield's departure. Not only that, Tryggvason went to Moscow repeatedly to support cosmonaut training and the first tests of MIM, all between 1994 and 1996.

Now Tryggvason, who was still patiently waiting for his turn on the shuttle, had a solid experiment idea to work on in space. Back in 1992, he and two senior Canadian officials in the space program (Aikenhead and Jill Sanders) established the basis of an agreement to have the CSA develop the Space Vision System further on another space shuttle flight, which would occur before the assembly of the International Space Station began, just to make sure the technology would be ready for the task. Although MIM had been in the works for the better part of a decade, Tryggvason said the timeline was still short to prepare for that flight opportunity, which was originally scheduled for July 1997 before slightly slipping to that August.

"We immediately started on the design and development of MIM-2, including updated electronics to improve performance over that of the MIM on the Mir," Tryggvason said. "The

MIM-2 flight hardware was manufactured by Routes Inc. in Ottawa. I also worked as an advisor to the scientists developing the various fluid science experiments. It all worked out with me doing a lot of commutes between the CSA in Montreal and JSC [Johnson Space Center] in Houston; to the CSA to oversee the hardware development and to Houston for mission training. A lot of excitement, a lot of work, but it all came together in time for the flight."

And in the best tradition of spaceflight, the MIM's legacy continued even after Tryggvason's flight. Following STS-85, he said, he proposed to the CSA that Canada should provide an isolation system to the experiment container in a fluid science laboratory developed for the European Columbus module on the ISS. The European Space Agency not only agreed to this but Tryggvason was appointed technical advisor.

His workload was extremely heavy in these years, as in 1998 Tryggvason and Thirsk both moved to Houston to take on mission specialist training, which demanded of astronauts the capability to do spacewalks and manage more space systems than the handful of experiments that Tryggvason supervised as a payload specialist in 1997.

"Each spacecraft had a stack of manuals about two feet thick to work through," Tryggvason recalled. "Many classroom hours, study evenings, simulator runs, tests, etc. kept everyone working hard." One perk of the process was the chance to jump in a T-38 jet trainer regularly to work on flight proficiency, which at least got him away from textbooks for a few hours.

While Tryggvason was still working on the European fluid isolation system, he received another opportunity to work at the Shuttle Avionics Integration Laboratory facility, affection-ately known as SAIL. This engineering simulator for the space shuttle had exactly the same type of hardware and software used in the space shuttle, accurate down to the wiring configu-ration in the cargo bay.

"This [configuration] ensured that time delays, and any electromagnetic interactions, would be replicated," Tryggvason said. Astronauts using SAIL would fly missions — mostly simulating the launch phase — repeatedly to make sure that the software and hardware systems would work adequately even in response to failures. One of his most memorable modes was the return-to-launch-site manoeuvre, which required the shuttle to do a quick turnaround in the first two minutes after liftoff. "I flew this profile a couple of times and based on these became confident that this very unusual manoeuvre would work," Tryggvason said. Fortunately, no one ever had to fly this type of a mission abort.

While Houston was a zone of professional success for Tryggvason, he found that his time there was difficult on him personally. His marriage broke up in 1992, while he was in training for the MacLean flight. "The split came with many challenges. I put in a lot of effort to visit my two kids often and regularly," Tryggvason said, adding that by 2000 they were in high school in Florida, and he wanted to give them some stronger Canadian connections.

His solution was to use a CSA program put in place to allow astronauts occasional study leaves, similar to a sabbatical in the university environment. While he was looking around for ideas, the CSA was approached by a Toronto-based company, Gedex, that had an interest in the electronics developed for Tryggvason's isolation systems.

Essentially, the big problem of isolation systems is this: they are tested in Earth gravity (also known as 1 g), but they require sensing of acceleration levels as small as one-millionth of the Earth's gravitational acceleration. This means that all the systems need to have two settings to work properly: one for testing on the ground in normal Earth gravity, and one for space operations.

Further, the space station was going to be a much more complicated environment than the space shuttle. Since it was larger and had more dynamics, the electronics would need to be set with a wider range to accommodate this new environment. The ISS might require an isolation system ranging from one one-thousandth of the Earth's gravitational acceleration to a tiny fraction of a millionth (specifically, 250 millionths of Earth gravity). It was a much higher demand than the shuttle, at a still incredible 35 millionths of Earth gravity. By the way, Canadian electronics — from the École de technologie supérieure in Montreal — helped power this huge range.

Gedex ended up being such a good fit for Tryggvason that he stayed three years in the role of chief technology officer, rather than working for a year with them as he had planned. In 2004, he returned to the CSA — but now his kids were at Canadian universities, making Tryggvason want to stay close to home. So in 2005, he started working at the University of Western Ontario as a professor, lecturing in space systems design and flight mechanics. By 2008, his mind was made up — he retired from the CSA to focus on teaching and on test flying; while he has changed teaching institutions a few times, he continues to work on these subjects to this day. He even flies airshows for the Jet Aircraft Museum in London, Ontario, still displaying the T-33 at the age of 75, and is doing flight testing on a new aircraft being developed in the US.

Changes in space policy came rapidly in the years after Challenger. The early space shuttle flights hauled a lot of commercial and military satellites into space; the United States decided to focus instead on using rockets for this purpose, which changed the nature of the shuttle program to be more scientifically focused. The idea for the Freedom space station ran into trouble as its budget soared to unimaginable proportions, forcing the United States to think again about how to construct a more affordable space station amenable to multiple international concerns.

The woes of the space station are so vast that they can (and have) filled entire books, but briefly speaking, the cost estimates produced by NASA were (as it turned out) nowhere near the true cost of launching, assembly and operations for Freedom. Space projects are notoriously hard to predict in terms of cost, due to the complications of working in a harsh environment with novel technology, but NASA has still come under criticism in the ensuing decades for underestimating the complexity of Freedom.

While NASA was discussing the costs on the US government side, the agency was signing on members of the international community to make contributions — including Canada, which was supposed to supply a dual-arm robotic system. Secretary of State for Science and Technology Tom Siddon, the minister responsible for the Canadian space program in 1983 and 1984, remembers signing the agreement in principle in spring 1985 to participate in the space station project. The partners would not only provide expertise, he said, but would also "give it some additional financial underpinnings."[5]

For Canada, a space station was part and parcel of its new space policy. While Siddon got the credit for it, he says much of the work was actually performed by Mary Jane Whiteman, a young science policy expert. She framed much of the policy, which included a call to create a Canadian space agency. That call still took a few years to achieve, but it heralded the beginning of the effort to centralize the various space interests in the country. A space station would give future opportunities for astronauts to fly, and help with the continuance of an astronaut program, Siddon said. Garneau and the rest of the group were still flying on a project basis at this time, so the hope was to pave the way for the program to become more permanent when CSA was ready.

"I might only add that as Minister of State for Science and Technology, my wife Patricia and I were invited to attend the launching of Canada's first astronaut, the Hon. Marc Garneau, in early October of 1984," Siddon said in comments as he was reading an early draft of this book.

"Prior to the launch we were honoured to attend — at 3 a.m.! — a private breakfast with Marc, the entire astronaut crew and Commander James Crippen of Mission 41-G, at the Kennedy Space Flight Center in Florida. What a glorious experience that was for my wife, together with the Hon. Marc Garneau. Several months later, in May of 1985, I was honoured to host the NASA director James Beggs under the Peace Tower in Ottawa, where he and I jointly signed the agreement in principle for Canada to become the first partner with the United States in the shared financing and construction of the International Space Station, which became the first tangible commitment by Canada to creation of the Canadian Space Agency, and getting Canadian's contribution to the ISS 'off the ground,' so to speak."

Canada was caught in some interesting international discussions in that era. The Reagan government was pushing forward the Strategic Defense Initiative (SDI) — an anti–missile attack system primarily created to repel a nuclear attack by the Soviet Union.

SDI was a delicate international conflict between the US and Soviet Union in which Canada was trying its best not to get involved. "Our government wanted to avoid being drawn into this thing that became known as Star Wars . . . but we didn't want to foreclose the tremendous opportunities for advancing the frontiers of science in space through the space program," Siddon recalled.

Meanwhile, the space station figures were becoming a difficult fiction to maintain. Howard McCurdy, a space historian who has written extensively about Space Station's Freedom's woes, said the cost discussion tied up talk in Congress for years:

"Every time they [NASA] would go back to Congress for a new appropriation for a new fiscal year, they'd have to work with these two figures. One that Congress was expecting, the other one was the real cost of fabricating the station. That absolutely paralyzed the station construction and planning process. To the point where NASA never got around to spending any money to actually build components of the space station."

MDA's Ower — who worked on some of the initial Canadian robotics concepts for the space station in the 1980s — said the robotic Mobile Base System proposed for the US space station is pretty similar to the Canadarm2 and Dextre that you will see today on ISS, with one exception: Dextre's central column used to have five joints on it instead of the one used today (for swivelling). That was so that Dextre could move in between the extensive truss structure used on the early space station. But the design of the space station changed iteratively as funding concerns mounted, and as the truss diminished there was no need for such a complicated design on the part of the Canadians.[6]

In the middle of this space station discussion, the new George H.W. Bush administration began tweaking SDI while deciding to launch a new human moon initiative in 1989. Announced on the 20th anniversary of the moon landing, the new moon landings were supposed to be an extension of the mission for Space Station Freedom (and perhaps, at the same time, make the station's cost more amenable to Congress.) By then, Truly — he who had flown the first flight with Canadarm — was NASA administrator and had a big fight on his hands when it came to budget priorities.

"That moon–Mars effort was politically the thing that we tried to do, and we failed," Truly said.[7] Some observers have said that the idea was doomed from the start, while others have suggested that NASA created very optimistic budgets for the moon program that would never be passed by Congress — especially because Bush was a Republican and Congress was a majority Democratic group. Bush's plan died quickly, and the space station idea came to be under threat, to the extent that Truly said he had to spend most of his time fighting for its survival.

It was not an easy period, especially because the space station concept in the United States came before Congress in 1992 and passed by a very narrow margin: just one vote. But it was clear that Freedom would not make it to space — it was too elaborate, too expensive, too difficult to launch in the tough fiscal environment of the early 1990s. Another solution was needed. The next chapter will explain how we got from there to the International Space Station, but suffice it to say, it required a new and unexpected partner — the Russians. Bringing the former Soviet Union into the fold led to a lot of policy changes, including shelving SDI and integrating launches and modules into a new international space station.

Meanwhile, the post-Challenger environment was a fruitful one for Canada, which eventually brokered an agreement that allowed Canadians to take an incredible run of one shuttle flight per year for several years. This run started in 1995 and ended in 2001, sending up Hadfield (twice), Garneau (twice), Thirsk, Tryggvason, Dave Williams and Julie Payette. Rather than focusing on each of the spaceflights themselves, a subject well covered in other sources, I am more interested in what we learned during this time.

One thing Canada learned was how to take on more responsibility in the space shuttle program. Starting with Garneau and Hadfield in 1992, every astronaut was assigned to mission specialist training. While the initial flights of Garneau, MacLean and Bondar were all payload specialist flights — their prime focus was science. They were responsible for the experiments that made up a portion of the mission objectives and a few other responsibilities like Earth observation, bailout training, environmental life support, etc. The outstanding work of the Canadians on these flights gave NASA the confidence to invite our astronauts into more advanced training. There also was a change in the greater environment after Challenger exploded, NASA astronaut Bolden recalled.

"We gathered our senses after we resumed flight again, and then started looking for ways to bring people in from the outside who would bring in some expertise that we may not have had on the inside. The big thing was looking at the international partners to fly, more so than trying to find American non-career astronauts," he said.[8] "I think it was largely, as I understand it, due to the fall of the Soviet Union, the belief by President [Bill] Clinton actually, that we had to find a way to keep Russian engineers and scientists from going to really bad places. And so that sort of began our efforts to try to find a way to engage with more internationals."

Senior leaders within NASA recognized the critical role of international collaboration in the success of the International Space Station. George Abbey, director of Johnson Space Center (JSC) from 1996 to 2001, foresaw the importance of NASA astronauts flying long-duration missions aboard the Russian Mir space station and used an inter-agency agreement between NASA and the CSA to appoint Dave Williams as the director of the Space and Life Sciences Directorate at JSC in 1998. Outside of the Avro Arrow appointments of Canadians in NASA, this was the first time a non-American held a senior executive role at NASA.

By 2001, Canadians had built up enough experience (among the group) to operate robotic arms and to perform spacewalks, while assuming more responsibilities in the shuttle hierarchy, such as troubleshooting systems issues while on board.

This run of flights also allowed Canadian technology to be developed quickly, since Canada had the flight credits and the astronaut time available to push forward their own experiments. For example, a version of the Space Vision System was used on Hadfield's STS-74 flight in 1995, when he connected a Russian docking module to Mir using the Canadarm.

This vision system helped Hadfield "see" the distance accurately between his targets. In space, where lighting conditions are inconsistent and it's hard to tell how far apart things are, SVS was used to attach the docking module between Mir and the space shuttle. More tests came as well: first Tryggvason's in 1997, and then Payette tested the Space Vision System again on STS-96 in 1999.

Another facet of this era was it showed Canada's emphasis on life sciences, which is another key theme in International Space Station research. Today, astronauts typically spend at least six months in space. Then they arrive back on Earth with a host of medical issues to contend with — weakened muscles, brittle bones, internal fluid changes — ailments that are normally confined to elder seniors. The hope is that by addressing these issues in space, we can better take care of seniors on Earth.

A few examples of this research: Thirsk's STS-78 in 1996 examined muscle atrophy and bone demineralization, Dave Williams's STS-90 in 1998 examined how astronauts spatially orient themselves in microgravity. The mission was called Neurolab and was NASA's contribution to the Decade of the Brain. Arguably the most complex life science mission of the Space Shuttle era, Neurolab also proved that international collaboration in space science worked. Science lead Mary Anne Frey oversaw investigator teams from nine countries with 31 experiments, two of which came from Canadian investigators, involving six space agencies as well as six institutes from the National Institutes of Health, National Science Foundation and the Office of Naval Research. The ground-breaking scientific results of the mission were published in a textbook outlining the many contributions to understanding the function of the central nervous system, while the operational lessons in scientific collaboration paved the way for research aboard the International Space Station.

In recognition of the unique scientific contribution of the mission, the crew requested that the Cajal Institute in Madrid provide microscope slides from the collection of Nobel Laureate Santiago Ramón Y Cajal to fly aboard Columbia with the crew, in recognition of the global contribution of neuroscientists to understanding the brain. Following the mission, on March 24, 1999, Williams returned the slides to the king of Spain at the Royal Palace of La Zarzuela in Madrid, where NASA and the crew received scientific recognition from the Consejo Superior de Investigaciones Científicas.

It was clear that the future of space exploration would be built on collaboration, and Canada showed that it was a worthy partner for space station construction. This not only included the space communications and training and mission support work that most people don't know about, but also the high-profile work that astronauts performed on the International Space Station in the later 1990s and early 2000s. During STS-97 in 2000, Garneau himself manoeuvred the Canadarm to install solar arrays on the International Space Station. These generate electricity for station systems including life support and science experiments. Then, Hadfield helped install the next generation Canadarm2 in 2001 during spacewalks on STS-100.

One could reasonably ask the question, so what? So what if about half a dozen Canadians got to fly into space during seven years? Is there any worth to us on the ground? In a few words, yes, there is. Besides the life science experiments and the industrial work already described here, there was also a community of researchers and engineers that got to participate.

For every experiment that a Canadian astronaut touched, there was a team on the ground — often a university team with students involved — that designed it to a standard worthy of peer review, convinced NASA to fly the thing and then analyzed the results (sometimes over many missions, taking many years).

When technology such as the Space Vision System or the Canadarm got to fly in space, this kept a huge team of engineers busy on the ground. It required decades of work to prepare the technology for space and keep supporting the technology during the spaceflight — a huge industrial complex that required companies to keep highly trained people on their teams.

You can also point to the intangibles, such as inspiration. These are always much harder to catalogue, although the Canadian Space Agency can point to metrics such as the number of students an astronaut speaks with, number of downloads on a website (or today, likes on social media), or number of submissions to an astronaut contest. Tracing how students follow the space program through the crucial years of university and early adulthood is not well tracked by any space agency, simply because the logistics are beyond them. So we work off of anecdotes.

"One of the reasons I wrote children's books was to capture the imagination of the net generation of the possibilities of where a passion for science, technology, engineering and mathematics might lead," said Williams after reviewing an early draft of this book. His memoir, *Defying Limits: Lessons From the Edge of the Universe*, speaks of the importance of resilience and commitment in overcoming the many hurdles that often arise in the pursuit of our dreams. "It is what we do when we don't succeed that determines if we will succeed," he said. Not bad advice for aspiring astronauts.

I know people who work regularly with astronauts, sometimes as reporters and sometimes as people who are directly involved in the missions. In most cases, when I speak to these folks, they remember following the program in childhood. They can identify individual astronaut flights that they watched.

And sometimes even politicians get inspired. I found this out when talking with John Manley, who (along with Mac Evans, whom you will meet in the next chapter) helped negotiate this sequence of flights in the 1990s on behalf of the Canadian government. Manley has an inherent interest in spaceflight and wanted to do his best in the halls of government to make it a priority.

"So you know how kids collect, like, hockey cards and baseball cards? There was a set of cards when I was a kid, like under 10, they were space cards. And there was a collection. I think there were about 80 of them. And the other kids were getting baseball cards, and I was getting space cards."[9]

These were the first words I recorded in April 2019 from Canada's former deputy prime minister and the former president and CEO of the Business Council of Canada. To hear this from Manley, somebody representing the highest echelons of government and business simultaneously, is a wonder in a world where, on most days, Canadian media relates few stories about any kind of science, let alone space exploration.

Manley, who was born in 1950, was 13 years old when John Glenn took his first orbital flight in 1962. Manley watched the coverage live on TV. His interest came full circle about 35 years later, when Manley was a minister responsible for the Canadian Space Agency and Glenn returned to space as a 77-year-old, later coming to Ottawa on a goodwill visit. He and Dave Williams, whose directorate oversaw the scientific research on the mission, discussed the importance of the Canadian bone loss experiments with members of parliament and Prime Minister Jean Chrétien. Glenn signed Manley's copy of Glenn's autobiography. "So I was very keen on him," Manley said.

Unfortunately, school depressed his interest in studying science — "that's run out of you often," Manley said — but the interest stayed with him. He won his first election for the Liberal Party in 1988; five years later, he had his first cabinet post as minister of industry. The year was 1993, and as Manley pointed out, it was not a good time to be asking for anything in the space program: "We were broke. Like, literally broke, to the extent that early in my term, at one point, we had a Canadian government bond auction that went no bid with 30 minutes to go. So we were right on the edge of the precipice of defaulting on our national obligations."

By 1995, his portfolio had lost half of its budget. And space was a vulnerable sector, because it had never been funded as an ongoing program. Manley says that, in his view, the program was overly politicized when Mulroney's government allotted a large share of money to a new headquarters in the quiet suburb of Longueuil, QC, which in the 1990s felt far from nearby Montreal — let alone Ottawa, where most of the space industry of the day was located. (Today, a toll highway road that happily swings around traffic-heavy Montreal makes Longueuil easier to access from Ottawa.)

Worse, Manley and CSA president Roland Doré had difficulties working with each other — both men expressed frustration to me about the other person in separate interviews. The timing couldn't be worse; in the 1994 budget, Canada was prepared to withdraw completely from the space station program. This wasn't Manley's idea, but an instruction he received from higher up. It took a phone call from US President Clinton to swing Canada's support back the night before the federal budget was released.

So Manley took charge. He asked Evans — who was running a company that coincidentally Manley had once incorporated as a lawyer — to negotiate on behalf of Canada to increase our astronaut flights in exchange for space station support. Evans and Manley together hammered out a long-duration plan for the Canadian space industry that included support for Earth observations with a radar satellite (Radarsat) to monitor climate change, and the creation of a next-generation successor to Canadarm (which ended up being the Canadarm2 robotic arm and Dextre robotic "hand," or manipulator, which helps set up equipment for spacewalks and does some exterior ISS maintenance).

Somewhere in there, an impressed Manley appointed Evans as president of the CSA. Doré moved to president of the International Space University, which is based in Strasbourg, France — happily separating the two by an ocean.

The Manley–Evans duo remains so much a part of Canadian space lore today that people repeatedly spoke their names, always in tones of awe, in my interviews — including Governor General Payette herself. Their apt negotiation skills, coupled with a genuine interest in making spaceflight work for our country, gave Canada a large set of shuttle flights to show our skills.

But the pace began to slow as Evans and Manley left their roles for other things, and NASA struggled with the early construction of the ISS. Sean O'Keefe, who was NASA administrator between 2001 and 2005, said he inherited a space station behind schedule and over budget: "As usual with anything of this scale, magnitude, and audacious size, it was something that was really a misunderstanding on the part of some of the participants in terms of what would be reasonably expected in terms of contributions as well as benefits that would be attained," he said.[10] These years had a lot of meetings between NASA and the international partners to determine what the sequence of missions would be, which caused some changes in funding and flight assignments behind the scenes. While the Canadian space program grappled with these changes, political priorities in our own country were changing.

By 2002, Allan Rock was minister of industry, and he lamented that repeated attempts to increase the CSA budget were not successful, even with a majority government to work with and even with astronaut Garneau heading the CSA and trying a push for a robotic Mars mission. The argument in government would go like this, he said: "[It] had to do with Canada becoming more present in the whole dimension of space, whether it was through a launching satellite, whether it was through purchasing a larger role in the International Space Station, whether it was developing innovations of our own."

It appears that after years of Manley and Evans trying to help Canada gain more prominence in space, the Jean Chrétien government was holding back further expansion. As Rock described it, "People said, 'Well, wait a second, we don't see that as a priority for the government. We don't know why we would put money into that. The Americans are leading, the Russians are up there. We're playing a supportive role and Canada's always made a contribution to the Canadarm, and we have astronauts, of course, but don't expect us to put our money up front. We're not going to compete with the Americans. There's no reason why we should.'"

It was an unfortunate development for Rock, a minister who (like Manley) described himself as enthusiastic about science — he followed the Mercury program (including the Glenn flight) when he was 13 to 14 years old, kept a close eye on the subsequent Gemini program, and remembers the Apollo moon missions some years later as "the absolute apex of excitement." And as minister of industry (as well as minister of health), Rock did what he could to leverage the expertise of astronauts he ran across, he said.

"I was thrilled when I met Roberta Bondar; I appointed her a science advisor when I was minister of health," he recalled. "I was thrilled to be on committees with Dave Williams. Of course, I've met Julie Payette on a number of occasions, before she became royalty. And just rubbing shoulders with these remarkable people was very exciting to me, and talking to them about space and the space program."

But despite his interest, he found that the CSA and its involvement in Canadian public life "were a relatively minor part of my work at Industry." He found himself caught up in the concerns of attracting more foreign direct investment, increasing entrepreneurship and helping start-ups go through growing pains while creating new ventures. Others in the industry department were very focused on putting high-speed broadband access into remote and rural areas of Canada, he added, which made it difficult to focus efforts elsewhere.

Rock argued that the Chrétien government was very responsive to science; between 1997 and 2003, it put about $11 billion to $12 billion into science and technology through the granting councils, created the Canadian Institutes of Health Research and started Genome Canada. But much of this was applied research that was easier to sell as a business case — space-related investments are usually for projects years or decades down the line, making them a more difficult sell.

Rock and Garneau did work together to create a strategy to try to make the CSA more attractive for the budget-conscious Department of Finance by "selling it as an integral sector of the economy," Rock said. They even tried moving outside the traditional aerospace sectors in Ontario and Quebec to the West Coast, where he had heard of a coalition trying to develop satellite components. "We just didn't get any traction in Ottawa," he lamented. "There was no interest in investing in the space agency at that time. And I'm afraid I disappointed Marc, and I was sorry about it myself, but it just wasn't [seen as] a big issue."

NASA's O'Keefe, however, said that Garneau was one of the most instrumental country leaders (along with the European Space Agency's director general) in keeping the space station partners in consensus about construction schedules,[11] but Garneau did this amid a low point in Canadian government interest. One reason for this disinterest may have been unfortunate timing: the Columbia space shuttle broke up Feb. 1, 2003 — right in the middle of Rock's two-year tenure.

O'Keefe, still NASA administrator when that accident occurred, was on the runway in Florida waiting to welcome his space shuttle crew back home. Instead, he found himself standing among family members, grappling with the loss while trying to be supportive of the space program. "It was a really horrific day; it was one that was seared in my memory forever," he said.[12] NASA put its space shuttle flights on pause until 2005 for the investigation, modifications and shuttle operations procedure changes, while doing what O'Keefe called "a soul search" (with its international partners) about "what we were trying to achieve with the exploration objectives."

Canadian flights — including an upcoming one by MacLean — were put on ice. By then, Canadian politics was shifting quickly. Rock's time as minister ended in December 2003; shortly after, the new prime minister, Paul Martin, called an election and found himself with a minority government. Canada, in fact, would have minority governments until 2011; while such governments are more accountable to the different parties, at times it makes it difficult for more focused legislation to get passed.

"The relationship . . . with the Canadian Space Agency has been exceptional," said Bolden, who became NASA administrator in 2009 while Stephen Harper led a minority Conservative government. "Prior to Prime Minister Trudeau [being elected in 2015] and the cabinet that he has now put in place, the difficulty was getting money for the Canadian Space Agency. It was [that] every director of the Canadian Space Agency struggled to get a usable budget."[13]

And to be fair to the CSA, it was also wrestling with new priorities by the Americans. Shortly after Bolden came in under President Barack Obama's administration, NASA was tasked with deciding whether it was worth it to pursue the Bush-era Constellation program to go back to the moon and on to Mars. NASA eventually ended up dropping Constellation due to cost and schedule concerns, which sent ricochets through the international partners. With such uncertainty plaguing the CSA's major spaceflight partner, funding concerns were bound to happen with the agency as well. And they did. Budgets remained flat for the early 2000s, even as Canadians were taking on a new phase to help construct the International Space Station.

CHAPTER 4

Glass ceilings and North Stars

> My attitude was more like, "It's too bad this is so important. It's too bad that we're not further along that it's a normal occurrence for a woman to go up on a space shuttle flight. It'll be a wonderful day when this isn't news."
>
> — Sally Ride, first American woman in space
> (NASA Oral History, Oct. 22, 2002)

Royalty called me suddenly one November afternoon as I organized forthcoming interviews for this book in a spreadsheet. "Hi, it's Julie from the Governor General's office," said a person I didn't know. "Do you have a minute?"

My parents used to bring me downtown in the 1980s, 1990s and 2000s every time the Queen and her consort, Prince Philip, visited Ottawa — continuing a tradition that my mother's parents did with her and her siblings in the 1950s and 1960s, when they were living in more distant Toronto. My grandparents were fervent royalists, and my genealogy-obsessed grandmother thought we were Loyalists, too — but the necessary paperwork was missing, never found in her active lifetime.

(She still is living in her late 90s, as of this writing, but living with dementia in a professional facility.)

So I was answering for my ancestors when I said, "Yes, I have a minute." As it turned out, Julie on the phone was the press officer for Governor General Julie Payette, with whom she coincidentally shares a first name. Of course, the Governor General represents the Queen in Canada, making this call exciting for someone who comes from a family full of royal admirers. But this Governor General also had particular relevance to my career. She flew in space. Twice.

Julie on the phone had called to see if I'd be available to do an interview in Kazakhstan, since Her Excellency would be attending the launch of Saint-Jacques. Her Excellency, fluent in five languages and an adept diplomat, happens to be friends of that astronaut's family. And she would help the family if something went wrong during the flight at any point between launch and landing.

I said yes, but time worked against us in Baikonur that December. I briefly spoke with Her Excellency once, in between her engagements, but she was busier than expected and only did one brief interview with reporters while I was otherwise occupied. But I live in Ottawa and Her Excellency does as well, so I kept emailing her office after returning home to discuss possible times.

In late February 2019 came the definitive response: "I would like to invite you for a meeting with Her Excellency," a representative of her office wrote. I received a standard Outlook calendar invite with a more cryptic location than usual: "HRX Study." I asked where that was. "I re-sent the invitation with the address: 1 Sussex Drive," came the answer.

So that is how the middle-class grandchild of royalty-mad railroad employees found herself pulling up to the guardhouse of stately Rideau Hall on a sunny morning in mid-March 2019. A highly decorated military escort hung my thrift-store coat for me and escorted me through a short hallway to a sitting room — the room the Queen prefers for visitors when in Ottawa, I heard later.

I only had half an hour, and there were so many matters of protocol to navigate first. Stand when Her Excellency enters the room, curtsy, let her speak first, let her decide where to sit, I kept reminding myself. I managed — perhaps awkwardly — all four in the moments after she swept into the room. But as I quickly discovered, protocol gets complicated when food or drink is involved.

A butler — steps behind the small entourage — offered me tea without mentioning if Her Excellency planned to partake. Her Excellency stayed silent, after a beat, and I took that as a sign to say yes. When the tea arrived, he placed both cups — as elegantly as possible — on a minute table, on either side of my gigantic tablet, which was recording our interview. Her Excellency (to my private horror) had to reach over my computing device to get her drink, but she gently refused my offer to move the tablet.

And so we proceeded, for 30 minutes, Her Excellency carefully focusing on the astronaut corps in general or the goal of the Canadian space program during her missions. She was opinionated, blunt and always fascinating. At interview's end, she generously granted me one more question beyond the allotted time, and then she was gone.

Canada's second woman in space — like Canada's first, Bondar — only had a short time in orbit. Both women have collected accolades in Canadian society despite the inherent sexism and obstacles that many women face in the upper echelons of science and engineering. But as I found out, despite their clear ability to lead and to inspire, discrimination had a terrible effect on at least one of these astronaut's careers.

Tales of female discrimination in the space program abound — for example, the "Mercury 13" women who underwent and passed the same psychological tests as the first astronauts, and the "hidden figures" who had to use separate bathrooms and take on the less glamorous jobs as "computers" in the early 1960s because they happened to be not only female but African-American. Also: the first astronauts were drawn from the military. That was good for passing security clearances, but it also precluded any female participation until much more recently. That's partly why the first American man flew in 1961, and scientist Sally Ride, the first American female astronaut, didn't get her chance until 1983. (Although, the first male American non-pilot flew in 1972.)

To be sure, we have come a long way since the 1960s, but female participation among the astronaut corps alone still lags far behind men. Only one Russian woman has ever visited the International Space Station since it began construction in 1998, and that's with the Russians flying at least two of their citizens in space four times a year. And most space crews in general — regardless of nationality — still have majority male composition.

NASA knows and continues to respond. It highlights female mentors. It conscientiously recruited 50 percent female astronaut candidates in the last two astronaut "classes" — 2015 and 2017 — so the hope is this will reduce the male–female disparity in future crews. And the agency is working in general to be more friendly to people of all genders, efforts which included (during a visit I made to Houston in 2017) a large pride flag flown near the entrance of astronaut central, the NASA Johnson Space Center.

But there are stumbles. NASA prominently advertised the first "all-female" spacewalk in 2019 during Saint-Jacques's flight, but they had to postpone the milestone when it turned out that the two women participating in Expedition 59 wouldn't fit into the available spacesuits at the same time.

By all rights, this should have been a victory for listening to women and for following safety protocol because NASA astronaut McClain herself made the call about the spacesuit fitting issue *while in space*. (She discovered a spacesuit she used during her first-ever spacewalk days before was too large for comfortable movement and asked for a smaller size for the second spacewalk.) Everyone on the ground listened to and agreed with McClain's reasoning.

But few paid attention to that. For interminable weeks, media headlines and social media participants cast NASA's decision as more discrimination. Never mind that Italian astronaut Parmitano nearly drowned in an American spacesuit six years before when Mission Control repeatedly minimized his calls of water in the helmet. NASA's safety practices were modified following an investigation. But few remembered the near-drowning; NASA was now damned as discriminatory, rather than praised for being more safety conscious.

The first all-female spacewalk finally happened in October 2019 after another spacesuit was sent up to the International Space Station, but in the meantime, there have been plenty of other female achievements to enjoy — a nearly year-long stay by NASA astronaut Christina Koch, NASA astronaut Peggy Whitson commanding the space station and shuttle commander Eileen Collins's historic return to flight mission after the loss of Columbia, among others. Canada's record-breaking female milestones are more modest, but with fewer astronaut slots available to us, all of the women we've sent to space have impressed.

Most Canadians of a certain age (myself included) vividly remember Bondar, who flew for eight days in space ("each one a jewel," one commenter remarked in reading an early draft of this book) in January 1992 aboard Mission STS-42. Remember that Bondar was selected along with a group of five other Canadian astronauts in 1983, so it had been a long wait for her. The vagaries of astronaut selection are often not discussed publicly, but factors that played into her wait include the Canadians focusing on their engineering missions first (the focus of Garneau and

MacLean) and also the demise of the crew aboard space shuttle Challenger in 1986, which grounded the fleet for two years and messed up the order of missions for ages afterwards.

Bondar is yet another example of how busy Canadian astronauts can be in between flights. In 1985, she was appointed chair of the Canadian Life Sciences Subcommittee for Space Station. This role came about due to a conversation Bondar had with then National Research Council president Larkin Kerwin, who passed away a few years ago. She argued that if there was preparatory engineering happening for the space station, the same should go for life sciences.

Canada, Kerwin and Bondar agreed, would be a good leader in this area not only because of our relatively healthy population in southern Canada, but also because of our understanding that the north needs access to health care, too. Kerwin recognized that Canada's strength in telemedicine positioned us to have a new role in the space station project, with respect to health care in space and post flight. "This is what you need for a space station. This what you need for the moon. This is what you need for Mars," Bondar said to me.[1]

In short order, she made a presentation to representatives of the major Canadian government research funding councils arguing for more funding for researchers in space medicine. It was still late morning on Jan. 28, 1986, when she made her way back to the Canadian astronaut office in Ottawa. "Of course, I walked back in the office and Ken [Money] was listening to the radio, and they said the Challenger's just exploded," she recalled.

Challenger — the same space shuttle that Garneau was on less than 18 months before — was destroyed just 73 seconds after launch, killing its crew of seven astronauts. Most people vividly remember the civilian teacher on board, McAuliffe, which was a painful loss on its own. But the incident also took down six NASA astronauts, a healthy mix of experience, newness and even cultural importance (as Ellison Onizuka was the first astronaut of Asian-American ancestry, and Judith Resnik had the mantle of second American woman in space).

One of the major ripple effects of Challenger — beyond the mission changes and design work — was that it changed how NASA did missions for the military and for astronomy. The United States gradually (over many years after Challenger) moved away from launching telescopes aboard the space shuttle, opting for expendable rockets instead (on cost and safety grounds). However, NASA still had a manifest of missions — most notably among them the Hubble Space Telescope and the Galileo spacecraft to Jupiter — that it needed to fulfill with a slower and safer flight pace, starting in 1988 when the shuttle was ready to fly again. So crews were reassigned to later dates to launch these probes in 1990 and 1989, respectively.

Also, payload specialists — the only class of astronauts that Canada was allowed to represent — was the category most badly affected by flight delays.[2] It was easier to justify flying a full mission specialist, who could do many tasks aboard the space shuttle, than an astronaut responsible for a small set of experiments. Bondar herself was selected for spaceflight only after a long (and more public than usual) competition with Canadian Ken Money for the biological sciences component on board a mission known as the International Microgravity Laboratory. A panel of 200 scientists from 13 countries picked Bondar over Money, which was reported as upsetting to Money as he was reaching retirement age and realized his chances of spaceflight were quickly diminishing.

The rumour went around that Bondar was selected over Money because she was a woman, and the community wanted to make a point of flying a woman in space — still a rare achievement in the late 1980s and early 1990s. Bondar didn't appreciate the insinuation.

"I mean, for crying out loud, I'm probably one of the most academically achieved persons that's ever gone into space,"

Bondar says today. (Her specialties in academic degrees alone include zoology, agriculture, experimental pathology, neurobiology and medicine — not to mention that when she was admitted as a Fellow of the Royal College of Physicians and Surgeons of Canada in 1981, that was in yet another specialty: neurology.)

Even after her selection, Bondar's flight date took a long time to be confirmed. Her launch was postponed an incredible 19 times.[3] The military had opted to stop the top-secret shuttle missions that it had run occasionally in the 1980s, but there was a backlog to finish after the shuttle resumed flights in 1988. NASA and the military ran seven joint classified missions even after the Challenger incident, with the last one going aloft in 1992.[4]

With little known about these missions, there are some wild yet unconfirmed tales awaiting declassification. For example, there's a persistent rumour that at least one of the spaceflights featured an unexpected and secret spacewalk in 1988, to fix a balky satellite antenna that refused to extend.[5] (How NASA managed to accomplish this with limited secure channels is another question.) If this is true, NASA will eventually have to revise the published order of spacewalks it so proudly discusses every time an astronaut crew goes outside.

In this post-Challenger era, with a backlog of missions to accomplish, orbiter switches were still common if it looked like a payload needed to be delayed. Bondar's mission itself was switched, compressing her planned flight timeline from 10 days to just 7.[6] The ongoing flight schedule shuffling also meant that Bondar flew ahead of MacLean, even though he was given a flight assignment in the mid-1980s and she not until 1990.

Bondar had been busy with the life sciences committee for years, undertaking cross-country flights and even a month-long sojourn in France to drum up support for life science research. She only stepped back in 1988 when, told she was in the running for a spaceflight, she was concerned about the perception of a conflict of interest in speaking to these people for life sciences when these same folks would be voting on her presence (or non-presence) on a future flight.

Bondar was determined to get into space even as a child. She enjoyed building plastic model rockets and speculates that if Canada hadn't opened an astronaut program, she would have moved to the US for more of her training; her grandfather was American, which could have eased the way. But she faced limitations. Women were not allowed in the Armed Forces at all, and their roles in the military were restricted, Bondar said. But she did end up flying and received her single engine fixed-wing rating as a private pilot as a young adult, while pursuing her passion "helping people and understanding how things work" in human physiology and biology. She found the dynamics of living organisms as compelling as human-made objects found in engineering and applied science, she said.[7]

The international astronauts, even when added to the flight manifest, were not welcomed with open arms in the 1980s and early 1990s in the least, Bondar said. "In the early days, the American astronauts did not want to have basically any internationals taking any of their spots away on the space shuttle and any of the prestige that they had," she recalled. This even extended (in some cases) to prohibitions in allowing astronauts on the flight deck, she says — the only spot on the space shuttle that had windows.

"I mean, it was ridiculous," Bondar said. "There were things that were totally unreasonable, but it was all a mindset to keep this image of a union type stuff together . . . There wasn't the same kind of respect and understanding that international people brought different types of energies and different types of viewpoints to a space program, even though space science for years has always been based on international work. Once you get human behaviour involved, then it's a much more emotionally based platform, I think. And I think that early on, [international] people were certainly not [the] respected crew members that they are now."

But Bondar's flight, she argued, helped show that international astronauts can do worthy work on missions. Hers was the first Canadian mission with an international component, she said; Garneau was responsible for Canadian experiments, while she worked on experiments from multiple countries. She said it gave the international astronauts "courage themselves to say, hey, we can do this." Additionally, NASA mission scientists and controllers organized the space shuttle crew into separate 12-hour shifts to simulate doing a longer mission in space and make up for the 10 day reduction to seven plus one, something that would preview time on the space station.

"It showed that we could all work together internationally," Bondar said of the various nations working on the space shuttle program, "and I think that really was the watershed moment for the space program in Canada. We've got a human being out there that represents Canada in the international community and is doing international work in space." Bondar had been told, she said, that the space program would never fly a female who was not American. Yet there she was. "Look at all the things we were able to accomplish by that flight. That really set a very good tone for Canada's involvement as an international partner in the human side of space flight.

The International Microgravity Laboratory mission, of which Bondar was a part, was able to perform 24 hours of science a day during a short period. Their experiments were breathtaking in their breadth, including matters such as how plants grow in microgravity, how ionized particles affect biological systems and (Bondar's contribution) how microgravity affects the brain and human orientation. The mission's work also was a step in helping with the ISS collaboration, including how to test modular laboratory hardware, instrumentation and stowage, and how to manage international partners. Besides generating dozens of peer-reviewed articles from the science, the mission was featured in the IMAX film, *Destiny in Space.*

Bondar returned to Earth on a public high, and did goodwill tours of Ottawa in 1992 that included a stop at the Canada Science and Technology Museum (where, incidentally, I saw her as a young child) and a stop at Parliament Hill, where she spoke with members of Parliament and had pictures with the sitting prime minister of the time, Mulroney.

"Then I took a taxi back to where the space program was situated, which was on Montreal Road," Bondar said. "I remember going in and all the secretaries said, 'Hey, great to see you.' And I went into the boardroom with one person who was an engineer, who was the head of the astronaut program at the time and basically he told me that my contract was over."

While at the peak of her public life, Bondar was told she needed to leave. Bondar didn't reveal this fact publicly for decades. Even her official CSA biography still stated, in May 2020, that she "left the Canadian Space Agency effective September 4, 1992, to pursue her research."[8] To say she was shocked on that day is an understatement. At that time, Bondar was the only neurologist who had ever flown in space. With Neurolab on the horizon, one early commenter on my book said Bondar would have been an obvious choice for that forthcoming mission. But that wasn't to be.

"'We'll keep you on for a year doing administration work, but that's all. You have to find yourself another job,'" Bondar recalls hearing. "And I was the only one who didn't say, 'Well, I'd like to know why?' I never did find out why. And I thought for years, I was thinking to myself, well, maybe they couldn't deal with two North Stars."

Bondar and Garneau were the only astronauts who had flown in space in mid-1992 (MacLean only got his chance in October). In the weeks and months after the incident, in fits of imposter syndrome, Bondar ruefully recalled Garneau's appeal to the media and his ability to speak eloquently (not to mention fluently in two languages). But she also remembers a firm belief that both males and females in the program "gives a good message."

"It was shocking, absolutely shocking that there was nothing . . . no one would do anything," Bondar added. "So I said, 'Well, I'm not going to do administration work. I've got all these university degrees now, this training, what are you thinking?'"

But Bondar was told she had no choice. She thought about ways to lobby to stay, but couldn't come up with anyone from her personal network that would support her. "Although I'm academically stellar, I didn't have a network, a political network," she said. "I wasn't born from parents that were in the Armed Forces, or any kind of diplomatic career, or lawyers. I knew no one.

"There was no one [who] protected me," she continued. "There was nobody coming to my defence and saying, 'Hey, what are you doing with this woman? She's just flown, it cost you $13–14 million [in 1992 dollars] to train her, and why are you doing this? Did she screw up? Did she kill somebody? Has she done something illegal? Like really?' And this has stayed with me for years. I never talked about it for years and years and years."

Bondar finally started talking about it in early 2018, because "I'm now 73, and I figure it's high time that people just stop this already." She said it has been difficult for decades to hear other people say she achieved her dream, because when it came to the old cliché of breaking the Plexiglas ceiling, "I had to use a blowtorch to get through it." Bondar points out that women in Canada have only (collectively) spent a few weeks in space, while men are approaching two years. She added that Sidey-Gibbons, the newest female Canadian astronaut, will have her own doors to pry open — although Payette and Bondar did their best to open doors before her.

"I've had discrimination against me as a woman throughout my life, in various careers. I never would have thought it would affect me then," Bondar said of her forced resignation. She wrote the letter of resignation, as requested, and said nothing about the circumstances publicly for decades, but it haunted her. "At the end [of the letter], I said, I wish I could have done more for my country, to represent my country, because when I was a little girl, I was just full of patriotism, but I couldn't join the Armed Forces. A little kid full of desires for these things, and perfectly healthy, and then is told, 'You can't do it because you're a girl.'"

Naturally, I brought this matter before NASA, the Canadian Space Agency and then-CSA president Roland Doré to see if they could shed any light on what happened. NASA said the situation happened so long ago that they no longer had anyone around who could comment. CSA read this chapter and offered no further commentary. As for Doré, his response was this: "After Mission STS-42 in January 1992, NASA decided that Roberta Bondar would not fly again on any USA space vehicle and consequently was withdrawn from the NASA astronaut program. CSA was informed of this decision. The vice-president responsible for the CSA astronaut program in 1992 knew the why. It seems that Roberta Bondar was not told the reason for NASA's decision."

Bondar, 48 years old when she resigned, still had many years of active work ahead of her. At the urging of a senior minister in the Ministry of Trade and Technology, she requested that the ministry would help fund some of her research work if it could be peer reviewed. "Of course, it was," Bondar said. "It's been published in many great international journals." The university supporting her, however, did not have or develop a mechanism to pay her as a professor, she sad. "I just couldn't afford to do it as a volunteer," she said. "So that's when I basically left the research."

So Bondar, the master of research diversification, diversified once again. She studied professional landscape photography in 1996 at the Brooks Institute of Photography in Santa Barbara, California. After having experienced the views of Earth from space, she photographed all of Canada's National Parks as they existed at the end of the last millennium as part of her personal project, Passionate Vision. Her passion for education, environmentalism and in particular her respect and love of birds (her Twitter feed refers to her sometimes as "RoBIRDa Bondar") are reflected in much of the work of the Roberta Bondar Foundation, which is aimed at connecting and reconnecting us to the natural world through the art, science and the technology of photography. She continues to inspire all generations to engage in a wide variety of sciences and the arts. She and Saint-Jacques collaborated during his space mission in 2018–19 to track migratory birds from space — a project that children could follow along with on their preferred devices.

The Canadian Space Agency experienced a lot of change in 1992. It was Canada 125, a year of national pride celebrating a major birthday since Confederation. The Queen came to Ottawa, a "Canada House" was set up downtown where Bryan Adams's best selling *Waking Up the Neighbours* album played, details of the new North American Free Trade Agreement came out and homosexuals were finally allowed in the military. And from a field of over 5,300 applicants, including 600 from children less than 10 years of age, the Canadian Space Agency successfully selected four new astronauts to join its corps: military test pilot and engineer Hadfield, aeronautical engineer Michael (Mike) McKay,[9] computer scientist and engineer Payette and emergency physician and neurophysiologist Williams.

While things were uncertain in the space program — the space station faced funding challenges in both the United States and Canada, as will be explored in the next chapter — the astronauts dove into their training with gusto.

Every astronaut class in Canada has a different feel to it, Payette explained to me when I asked about her own astronaut selection. Nobody selected in 1992 had a PhD, for example, because they weren't supposed to be research astronauts, she said. But in 1983, just about everybody did because their specialty was supposed to be monitoring experiments.

During our interview in Rideau Hall over tea, Payette didn't mention the challenges of being a woman in orbit. But Payette has always been a supreme achiever, confident in her work — whether she was singing professionally or working professionally. And like any good astronaut, she puts her role in the context of the greater team.

In our discussion, whenever I asked a more personal question, she would answer in terms of the team, the crew, the generic way things are done, rather than focusing on her own achievements (numerous as they are). And while her job today is diplomacy, she's still very much immersed in space. Not only is she friends with Saint-Jacques and his family, but she brings space constantly to Rideau Hall. One Hallowe'en, for example, she greeted trick-or-treaters in her astronaut spacesuit.

Payette waited four years for her mission specialist training, but she told me there was much to do to get ready for that training. Her background is in what today we would call artificial intelligence, particularly in the fields of speech recognition and natural language processing. Decades later, AI is partly responsible for why Apple's computer virtual assistant Siri can understand most of your shouted commands while you're driving.

Fundamentally, Payette was worried about system usability. Like everyone in the 1960s, she had grown up with computers that did their thing behind impenetrable screens of code; the idea of computer mice, of graphics, of clear instructions didn't become popular until the 1980s or so. The Apollo astronauts famously punched in sequences of obscure numbers into their space computers to make their way to the moon and back; the first landing on the moon in 1969 was almost aborted because an obscure "1202" computer overload error popped up in front of astronauts Armstrong and Aldrin, who had no idea what it meant. (Luckily, Mission Control did.)

These early computers got us to the moon and back, but the computers of the 1990s were more capable. Many houses and schools had personal computers available, with early versions of Windows or the Classic Mac OS running. And at the space program, the new class of astronauts helped older engineers come up with better ways to interface with the space shuttle. Payette's job was to help future crew members work with the Canadarm2, Dextre and the Mobile Base System then known as the mobile servicing system.

"They use crew members as advisors to develop the interface and how it works, and we go through a number of simulations and work with them," Payette said of her job in 1992. "We work with them, the operators. We work with them so that their system is usable. And that the procedure they write to use the system is understandable. That when they say 'this switch,' well, that switch is on the panel somewhere."

Payette was part of a network of crew interface personnel. You can't beat her description of how the job worked: "Anything that the crew will touch [or] use, procedures, switches, interface, computer, manipulation, look and feel — tap, tap, tap — and then, of course, operation." In a complex place such as the space station, this thinking is supposed to make the crew's lives easier as they go through procedures as routine as unclogging a toilet to as complex as preparing for a spacewalk. These easy-to-understand controls and procedures were meant to help the astronaut to figure out what's needed without memorizing long technical matters or calling home for help. But every engineering process takes a long time, she observed. Remember: this is 1992. Canadarm2 didn't take flight until nearly a decade later.

Payette felt it made sense to send Hadfield and Williams before her for mission specialist training, she said, as they were each older anyway. So, while still in Canada, she immersed herself in all she could to get ready for Houston. She learned Russian, she picked up a private pilot's licence, she became a parachutist, and she spent 120 hours as an operator aboard reduced-gravity flights — an incredible 60 or more times aboard the infamous "vomit comet," as NASA calls its airplanes that do up-and-down motions to simulate microgravity. "The entire time that I spent here in Canada was a very lucky time for me, because by the time I showed up at NASA, I had all of this helping me," she told me.

Payette had three years of training in Houston before her first flight in 1999, and the importance of that time on the ground should not be understated, she said — nor the support of tens of thousands of "very professional, very rigorous" people that helped the astronauts achieve flight safely. A single example of her work is the Shuttle Assessment Interface Laboratory (SAIL), which allowed astronauts to simulate a real flight — right down to the flight software astronauts used, which was Payette's particular interest. Astronauts, engineers and indeed an entire hierarchy of personnel made modifications to the software and tested it ahead of flight.

And then there were the folks who worked on the space shuttle themselves — people who, unlike much of the adoring public Payette met, looked at the astronauts as equals. "I have always been awed by the ownership that these people have," Payette said. "We would go and meet metalworkers that would bolt the big bolts in the skirt of the main engines of the space shuttle. That's what they do for a living. And when you meet them, it's not 'Oh my God. An astronaut.' No. It's 'Which engine are you flying on? Which mission exactly?' So you tell them, 'STS-127.' 'Ah, that's engine 265. I worked on that.'"

Payette paused to emphasize that last phrase. "They *worked* on that. They put the bolts on the skirt of that engine that you are so lucky to fly because 'Me, the professional metalworker, I did a really good job on that engine. I always do a really good job. Very happy to meet you, by the way.' 'Very happy to meet you, monsieur.' And that's something that is little known."

Payette was assigned a whole suite of tasks as a mission specialist during her first flight, STS-96, in 1999 — in seven years, already much had changed since Bondar's flight. Every Canadian astronaut flying in those days had the higher designation of mission specialist, and Payette came right in the middle of a "golden age" of Canadian flights that allowed one person to fly per year. She was designated robotics operator, an engineer for the Russian module, and a translator, because she was the only person on the crew who could speak both English and Russian fluently.

Payette emphasized the team aspect of her training and how they dealt with failures. "What do we do if the system fails? On the robotic system [for example], they will, in the simulator, introduce faults, and problems, and emergencies so that we will respond. At first, they introduce very few [failures]. One, find the procedure. You don't know where it is. The mission control helps you. Instructors help you. Your crew members that have gone before help you, but at the very end before you fly, you better know what you're doing. So, it's a gradation. The entire training during mission training is geared to work that."

Living on station would have meant an even more rigorous training process, Payette pointed out, saying that crew members there need to be everything from "the Maytag repairman" to operators, scientific proxies, cooks and cleaning persons. That's why it takes two to two and a half years for a typical crew to get ready for a space mission. "You train on the vehicle that you will launch in, come back in. You train on the systems of the space station, and spacewalks, and robotics, and language training — and all of this at the same time, while keeping your reflexes alive in simulators and aircraft."

After months of training for her space mission, Payette said the environment in space felt familiar — but there are things that could not be simulated on the ground exactly. Take robotics, for example — Payette operated Canadarm in space. The procedures were identical to the simulation, but the real thing felt different. "If I hit the side of the space shuttle in the simulator, I don't look very good, but there's no consequence," she said. "I didn't break a space shuttle, which is my re-entry vehicle. But if I do that in space, the margin for error is a lot smaller — if not non-existent."

If that doesn't sound complicated enough, there were simulators to match different aspects of working with Canadarm. There was an entire simulator devoted to grappling objects in space, where astronauts learn how to match the rotation and movements of a free-floating object, she said. This had stymied NASA back in the 1960s when they were trying to get two spacecraft to dock together. But as the space station came together, retrieving objects had to be routine. There was no time for error with only three to four shuttle flights a year and construction expected to take more than a decade if all went well.

Payette actually had two visits to the space station — one at the very beginning of its construction in 1999 (she was one of the first people ever to float inside the station while it was in space) and again in 2009, as the space station was in its final stages of construction. She met with Thirsk in orbit, making it the first time that two Canadians worked together in space. But there was little time to celebrate the cultural milestone, as Payette and her STS-127 crew were finishing construction of the Japanese Kibo module. This meant, incidentally, that Payette operated three robotic arms in space: the shuttle's Canadarm, the space station's Canadarm2 and a special-purpose Japanese arm on Kibo.

"I thought I was very privileged," Payette said of going up a second time. "Very privileged that I saw the station at the very beginning and I saw the station at the very end, with people on board on a permanent basis. So that was very rewarding from that point of view. I was also present when the first crew launched from Baikonur, and I was present when the first module launched, Zarya, in 1998."

Then she shifted her focus once again to the team, including the 1996 class of US mission specialists that she joined as a young and talented Canadian. "It's been a tremendous privilege for all of us that have been [working] during the construction missions. All my class of astronauts, we have flown the construction missions. All of us."

I've skipped over many of Payette's achievements in space, which have been well-documented. Indeed, within a half-hour conversation, there was not much time to talk about her milestones. Near the end of our discussion, however, she reached for a small pile of papers she had brought with her into the room. At the top was a journal from the Wilson Center, a non-partisan organization that, according to its website, examines "global issues through independent research and open dialogue to inform actionable ideas for the policy community."[10]

While still a Canadian astronaut, Payette took the unusual step of joining the Center for a project on the challenges of diversity in science education. She examined this issue for six months, between December 2010 and August 2011. And the booklet Payette held in her hands was one of the products of this educational immersion, this policy opportunity, that continues to inform her work as Governor General. Today, she not only performs the traditional ceremonial duties of her office, but she continues to promote science and space education where she can — including handing out awards for the Natural Sciences and Engineering Research Council, hosting a Canadian Science Policy Conference in 2017 and even handing out pictures of our planet during her various international visits.

Payette's tenure has not been without controversy[11] — with matters such as high office turnover and her reluctance to live in Rideau Hall until it's renovated.[12] She's always been intensely private, especially for the sake of her son. He was rarely photographed or even mentioned while she was in the public eye as an astronaut. Years later, when a group of Canadian media outlets went to court asking for her divorce proceedings ahead of her entering public office, her justification for refusing to release them was she wanted to shield her son. Payette dropped her challenge to releasing the proceedings, and the media group said their interest was due to her becoming the Queen's representative — not anything due to her son.[13]

All astronauts face this struggle, no matter what gender they are. We remember the Apollo astronauts for their landings on the moon and their excellent technical work, but many of these people divorced in the wake of the strain. Decades later, in 2007, space shuttle astronauts Lisa Nowak and William Oefelein left NASA in the wake of a public and messy love triangle. NASA swiftly implemented several measures after the incident, including an astronaut code of conduct and a promise to review its psychological screening.[14]

We remember astronauts as heroes, but we can never imagine the strain that heroism produces and the personal sacrifices they have to make. Everyone does their best to support the families and the kids, and NASA has many psychological countermeasures available to astronauts — including the ability to reduce somebody's schedule in space if that person feels under strain. But there still are difficulties. And after interviewing Payette and Bondar, I wonder if we — as a society — are doing enough to support retired astronauts in the years after their flights.

CHAPTER 5

Canadarm, cuff links and collaboration

> You know the greatest danger facing us is ourselves,
> and irrational fear of the unknown. There is no such
> thing as the unknown. Only things temporarily hidden,
> temporarily not understood.
>
> — James T. Kirk in *Star Trek's* "The Corbomite Maneuver"
> (1966), written by Jerry Sohl

While trying to make tough decisions, Chris Hadfield has some questions he tries to consider: What should we focus on? What's important? Where do we put our energies? And if you're making a decision in a position of responsibility, how does it affect other people?

In his life, this set of questions has allowed him to do many things — everything from deciding not to shoot down some Soviet jets in Canadian airspace, to leading an International Space Station crew in fixing an emergency ammonia leak, to figuring out how to get into the Mir Space Station with a penknife.

Hadfield is the only Canadian astronaut who visited two space stations, and perhaps more than any other, he symbolizes the transition from tension to peace between the Soviet Union and the Western world. He started his career when everyone worried openly about the nuclear threat. Yet shortly after the Soviet Union dissolved, he found himself on one of their space stations, picking up lessons in collaboration that he would carry for the rest of his life.

Growing up, Hadfield didn't see the Soviets necessarily as an enemy. In between his reading of science fiction and comic books, he watched the news and was fascinated by the space race between the Soviet Union and the United States.

"[The two superpowers] aggressively took this new technology and deliberately decided to get out in front of it and say, 'We're not going to wait for someone else to push this technology to the limits. We are going to lead. We are going to take control of this. We are going to be the ones,'" Hadfield said,[1] adding that world leaders' decisions at that time were both uncommon and necessary. While only two countries had direct access to space in the early 1960s, what these two leading countries did "had worldwide consequences and pushed back our levels of ignorance and allowed us access," he said.

Hadfield, of course, is neither Soviet nor American. As a 10-year-old Canadian who got inspired by Neil Armstrong's first steps on the moon in 1969, he was old enough to know his country had no space agency, no spaceships and a nascent space program that used American rockets. Looking back, though, he says it's clear that the Americans and Soviets wouldn't stay by themselves in space. "There's a whole bunch of stuff here that's going to take some time, but it is not going to be irreversible. You can't get all this stuff back into the bag," he said.

If Canada had turned its back on human spaceflight, Hadfield said he probably would have become an American: "I would've gone to the only place that offered the opportunity to do the things that I thought were important and worthwhile." He said maybe he would have been an unsuccessful astronaut wannabe, if he had had to apply through NASA instead of the Canadian

Space Agency. But that wouldn't have mattered. He wanted to be a productive citizen, to follow his interests, to have a government supporting "the aspirational, inspirational, enabling, self-visioning part of leadership and of decision making."

Luckily for Canada, Hadfield did become an astronaut. He toiled for 20 years and completed two space missions before he burst onto the world stage as the guitar-playing, tweeting, ever-humorous commander of the International Space Station. But to Hadfield, Expedition 34/35 was the latest logical step in a long line of decisions he made to benefit others — decisions he continues to make today even after retiring as an astronaut.

"It's why I wrote that book," he explains — that would be the bestselling *An Astronaut's Guide to Life on Earth*, a charming autobiography that explains how to think like an astronaut. It's also why Hadfield hosted a National Geographic show about planet Earth (*One Strange Rock*), why he ran a lecture series on astronaut work for online video-learning provider MasterClass, why he teaches at the University of Waterloo and so much more. "[It] is in the continued pursuit of the things that are important to me, and as my ability to be less selfish . . . my ability to maybe have a wider impact — or share it — has increased with time," he explained.

Hadfield is now well-known and recognized, and he makes the most of it to help other people. But Hadfield wouldn't be the Hadfield we know without people on the ground who picked him (and three others) out of more than 5,000 astronaut applicants and cleared the way for him to fly in space. Most of us will never know their names, but here's one you should definitely remember: Mac Evans. So powerful a force is Evans that decades after his retirement as Canadian Space Agency president, he still has the ear of politicians today. The space community universally speaks about him with respect.

The two men, both of whom lived in Sarnia for a time, spoke to me within the same week in January 2019. With their interviews coming so close to each other, I couldn't help but draw analogies — if Evans had been the younger one, and Hadfield the older, could they have traded places? Because they both have the same drive, the same desire to help others. And both, in their own ways, have had immeasurable effect on the Canadian space program as we know it today.

Evans started his career at the Defence Research Telecommunications Establishment, the same location where Alouette and the early satellites were built. This department (now part of the Communications Research Centre) tested a lot of the technology that is in today's computers, from a modest building in Ottawa's west end, just down the road from ill-fated Nortel. Evans said he went there out of an interest in space. Like Hadfield, "I wanted to go where the program was. In those days, that's where it was," he told me over tea in Ottawa.[2]

Evans joined in 1970 and worked on Hermes, the largest and most powerful communications satellite in the world when it was launched. Today, direct-to-home TV broadcasts are a service we take for granted — it was Hermes that pioneered this technology. Evans, a 30 something near the beginning of his career, was mission director for the program, which he characterized as "probably the best job in the world."

But the call to public service was stronger for Evans. So when Hermes became operational in 1976, he moved to the policy side and worked two levels down from the mighty Chapman, who had made the first survey of Canada's space industry years before. For three years, until Chapman's sudden death in 1979, Evans worked in the Industrial Development Group at the Department of Communications to make a space program supportive of Canadian industry.

Chapman compared space to the old railroad industry in Canada, believing that the trains — or spacecraft — could only run/fly with the support of industry. So the team did things such as negotiating with the European Space Agency to get Canadian industry into their labs and working with NASA to get more satellites on their rockets.

What's important to understand here is that Canadian space research and activities were not only conducted at Chapman's lab. Earth observation flowed through Energy, Mines and Resources. The National Research Council continued its pioneering space science research. The Department of National Defence also had an interest in space, although it was "very little" in those days. Eventually, these various parties realized that their interests should be united, and they created the Interdepartmental Committee on Space, chaired by none other than Chapman (then the assistant deputy minister).

But space was shifting quickly. In 1975, future Governor General Sauvé, then minister of state for science and technology, issued Canada's first space policy, a high-level document that set just enough direction for the engineers and scientists to get going. And NASA was itching for international participation in the space shuttle program and was asking for ideas from Canada as soon as 1974. Evans, sensing an opportunity, jumped over to the Ministry of State of Science and Technology (MOSST) instead. It turned out to be a prescient move. More programs began flowing through the ministry, and before long, MOSST was chairing the Interdepartmental Committee on Space. MOSST's programs included name-brand projects that still echo in Canadian space circles today, 40 years on: for example, Radarsat and the predecessor work to what we now know as the International Space Station. MOSST also, eventually, ended up creating the Canadian Space Agency — which "focused the government's attention on space over an extended period of time," Evans said.

The CSA was founded on March 1, 1989, at the height of the space shuttle program and while NASA was still trying to figure out how to create a permanent outpost in Earth orbit, called Freedom. The Soviets had a space station — it was called Mir — but with rhetoric flying from Reagan's office about the nuclear threat, nobody imagined any Americans setting foot (so to speak) in Mir. So there were not only competing countries, but also competing space stations — a theoretical one to be built with international collaboration and a functioning one under solely Soviet purview.

Those who participated in Freedom would be guaranteed flights, with the number of flights tied to the contributions they made. Everybody expected Freedom would be ready in the 1990s, so Canada held an astronaut selection and found four new people to join the corps. Everybody expected Canada would take on more responsibility in the space station, so Garneau and Hadfield both went for mission specialist training shortly after their selections.

But Freedom, as explained in a previous chapter, was in big trouble and came within a vote of being killed by the United States Congress. That's well-documented. Lesser known is that Canada had its own woes as well, because it was running into cost overruns and schedule delays on the mobile servicing system (which would become Canadarm2). In the tough fiscal environment of the time, a project with no clear end in sight wasn't pleasing to the politicians. So, in the 1993 budget, Canada wrote up that it would withdraw from the International Space Station. It took a last-minute intervention from US president Clinton, who personally negotiated with Canadian prime minister Chrétien, to keep the space station going.

But Canada needed a plan and it needed it fast. Evans — like many public servants of the era — had left for private industry, as it seemed a safer haven. The minister then responsible for the CSA, John Manley, needed a champion. He needed someone who knew the space program inside and out, but who also wasn't currently under government employment, Evans explained. So as Manley worked to put together a space plan to stop Canada's space program from undergoing further cuts, he hired Evans. The two men worked closely together for three months brokering a deal with NASA, which was pushing harder than ever for international collaboration in the wake of the 1992 vote.

"They were desperately needing us to stay. We felt we could stay," Evans said. In 90 short days, they found US funding for Radarsat, made a successful case for creating Canada's first science satellite in 30 years and promised a redeveloped Canadarm. This work formed the basis for all the opportunities Canadian astronauts had to fly in space between 1995 and 2001.

Hadfield had had his eye on the Canadian space program as far back as the 1983 astronaut selection, which he didn't apply for because he only had an undergraduate degree at the time. But he watched these space flyers closely and realized that he had a choice about whether to try for astronaut-hood: "The door got kicked forever open, and it was now up to me to decide if I could walk through it or not. So that was a real 'put your money where your mouth is' thing. Okay, big talker, you've been saying you were going to be an astronaut since you were nine. Now you can. What are you going to do?"

Every astronaut selected is at the top of their field, and then they get kicked down the ranks into rookie-hood the moment they accept their assignments with the CSA. For Hadfield, the change was more jarring than for most, because he was one of the first Canadian astronauts to go directly to NASA. He shared an office with Norm Thagard — who would be on his way to Mir shortly, as the first American there — and John Young, the moonwalker who flew the very first Gemini and shuttle missions, and was chief of the astronaut office for more than a decade.

"It was pretty easy to drop into the role of being a tiny, little fish in a brand-new, huge and voracious pond," Hadfield said. "But it was just inspiring to me. It's like, 'Holy crap. Everything I've done so far has just been buying me the price of entry, and now the real question is, can I keep up? All these things I've been trying to do that got me in the door, do they matter? Were they enough?'"

Still Hadfield was very much reminded of his place in the rankings. He was an astronaut candidate — ASCAN in NASA parlance, "which couldn't be more derogatory," he said.

While astronaut candidates are hailed as heroes when they stop by their hometowns, at the Johnson Space Center they very quickly realize "you're an ASCAN who knows nothing, and you're responsible for the Christmas skit." But Hadfield embraced the chance to learn and to move up. After all, he reasoned, "Norm and John were ASCANs at one point, and they had to earn their stripes. Nobody handed it to them, so to me I just found it humbling where you needed humility. So that's all right."

Hadfield also needed to take on a new persona. Fighting the Soviets had been his raison d'être in the 1980s. Time-travel to 1992, as an ASCAN, he very quickly realized that he was part of a larger effort to think differently. At NASA, and in the United States, the powers that be were fighting to keep Soviet rocket technology from "spreading to the four winds" while the newly formed Russia struggled to get its feet underneath it. While Hadfield had nothing to do with that effort, he listened. And learned. When he got the chance to take Russian lessons, he took it.

It felt strange back then, he allowed. "Sitting in a classroom with this young Russian instructor with a bunch of other astronauts, some of whom were taught from a young age to hate the Russians, to distrust them and view them as an evil, malevolent blackness just beyond the horizon. And now having to learn the language."

But Hadfield knew — as did the other astronauts and ASCANs — that the United States and Soviet Union had run the joint Apollo-Soyuz mission in 1975, and as Hadfield's knowledge of the language deepened, he got to know those participants. That included Russia's Alexei Leonov. If the Soviet moon technology had been up to snuff, Leonov would have been the first to walk on the moon, Hadfield said.

Hadfield met Leonov in Russia when Hadfield found himself assigned to Mir — on a space shuttle flight, STS-74, that would add a new docking port to the space station to make it easier to do shuttle flights. Hadfield was assigned to STS-74 probably due to engineering and test pilot qualifications and experience. He said he was the best Russian speaker of the crew, and despite being new to the language, he was working fast to get to know the culture and the people.

His voyage to Mir is well covered elsewhere, but it is worth telling the moment when Hadfield had to "break into" the space station upon docking. Hadfield found himself with an impenetrable hatch when the space shuttle docked with Mir, and one of his tasks as junior crew member was to get everybody inside.

"Some very strong and well-meaning technician from Russia had decided that his part of the station was going to be secure, and he had strapped down that hatch and put a webbing across it, and wire ties. It was almost impenetrable to get through this hatch," Hadfield said in one of his MasterClass sessions. Thinking quickly, Hadfield grabbed a jackknife and swiftly sawed through the cables and ties. "I felt like I was breaking into the Russian space station all by myself," he joked. "The moral of the story is, when you're going into somebody else's spaceship, bring a jackknife."[3]

Speaking with me, Hadfield said using the Canadarm to put a fresh docking module onto the space station — to physically allow the space shuttle to dock with the Mir, without having to move entire modules of Mir like they had to for STS-71, was something that he would remember for the rest of his life. "Looking back and seeing our docking module attach to Mir, it was phenomenal and amazing, and to me, way better than intercepting Soviet Bear bombers off the coast of Newfoundland."

The decision to fly American astronauts to Mir was not without controversy, although NASA senior officials such as George Abbey embraced the opportunity for many operational reasons, including a redundancy in launch capability should something happen to the space shuttle. Famously, the aging space station experienced a series of near-catastrophic events while Americans were on board — most notoriously, the time a cargo ship slammed into the spacecraft and the time a fire broke out on board (in a place that, astronaut Jerry Linenger claimed, had faulty smoke masks). I remember opening an *Ottawa Citizen* issue around this era and seeing a proclamation by Lovell — he who survived the Apollo 13 explosion while en route to the moon in 1970 — saying the Americans needed to come home fast.

The American astronauts and Russian cosmonauts battled cultural issues, even over small matters. Astronaut Bolden, whose wife is a realtor, said they remember getting complaints from a homeowners' association after placing two cosmonauts in houses in Houston.

"The guys wouldn't cut their grass, and they had weeds in the yard. And we couldn't understand it, so one of the ways around it [was] we just hired yard maintenance crews to come in and take care of that for them," Bolden recalled. It wasn't until after he had flown a mission with the Russians and went to visit them after his flight that he saw that every apartment in downtown Moscow had weeds in the yard. "They had this incredible appreciation for any kind of plant life, and they wouldn't think about cutting it," he said. "They just thought that was how you were supposed to do it. They didn't realize that manicured lawns and stuff was a way of life in the United States. So, it was just a cultural thing."[4]

To be sure, NASA carefully investigated its options mission after mission. It talked to its astronauts. It talked to Congress. And more tellingly, it kept talking to the Russians as well. The two agencies had very different styles of communication due to different cultural experiences. Many Russian astronauts speak of having to memorize procedures without access to textbooks, for example, because there was a shortage of paper in the Soviet Union and in Russia while the space program was growing up.

Mir was a risky proving ground for American–Russian collaboration, but it was a valuable one nonetheless. The United States kept pushing the line that it needed that experience to collaborate on the International Space Station, which was starting up around the same time as the Shuttle-Mir program put American astronauts on Mir.

To support Mir missions and the nascent ISS program, Americans found themselves living in the newly opened Russia and learning what daily life was like there. NASA oral transcripts from that era relate a wealth of interesting experiences that, really, you could write a whole book about. A typical account — Melanie Saunders, who worked with NASA on export control, described one of the logistical challenges the agencies faced.

"Export control regulations had not really caught up with politics and world events, and the Russians were still bad guys. So it would be sort of like — imagine if things changed with North Korea, or as things have changed with Cuba in recent months," she said in 2015. "Suddenly you're trying to exchange technical data, and according to all the export laws and regulations, they're still the bad guys. So there were a lot of hurdles to us being able to just get together and do technical interchange meetings and such."[5]

There were massive problems to overcome beyond export control regulations. One of them concerned the launch of Zvezda, the second Russian space station module for ISS. A lack of funding in Russia pushed back the initial launch date of April 1998, but a chain of events delayed it further to July 2000. The reasons are complicated — as with any big engineering project. A Proton rocket of the type expected to carry Zvezda into space exploded, necessitating an investigation. Technical issues with the space shuttle program also contributed to the delay.

But the climate was (at least in public eyes) quite hostile, with NASA administrator Dan Goldin accusing Russia of "dragging its feet" in February 2000 and threatening to throw it out of the ISS program altogether if Zvezda was not launched promptly.[6] What was lesser known publicly was the Americans had delayed modules as well — and that perhaps the real impetus behind the comments was American frustration that the Russians continued to deploy resources to Mir[7] up until its controlled deorbit and demise on March 23, 2001.

It's difficult to imagine any multinational project, no less a multinational project between former Cold War rivals, arriving to fruition without hiccups. Indeed, during Hadfield's next mission in April 2001, the coalition faced another potentially pricey delay. Hadfield — who was supposed to be installing the robotic Canadarm2 that was designed to heft bulky space station components for assembly — found himself blinded during a spacewalk. At the time, nobody knew why, but later it was determined that some cleaning fluid from Hadfield's helmet glass got into his eyes.

Hadfield chastised himself harshly for this in *An Astronaut's Guide*. His spacewalking crewmate, Scott Parazynski, later told me that — naturally — NASA had a procedure available if Hadfield became disabled. "I was very close to having to do a what we call in class incapacitated crew rescue and bring him into the airlock, but he was able to, eventually, generate enough tears to flush out the soap that had gotten in his eyes," Parazynski recalled.[8]

Let's remember, though, that Hadfield was no space assembly rookie. He may have been taking his first two spacewalks, but behind the scenes he had been a CAPCOM for many years after getting back from Mir. ("Suddenly, I had a lot more to contribute" to being CAPCOM, he said of his experience after flying once in space. Hadfield worked 25 shuttle flights in a row, "through the pivotal early space station assembly," he told me. A lot of this work included building relations with the international partners, particularly the Russians, through these tricky first space station assembly flights.

Perhaps NASA recognized that work, because Hadfield — oddly, to his mind — was made the lead spacewalker on his first spacewalk. That's a very rare feat, and Hadfield was still a young astronaut at only 40 years old. He remembers spending his time before the flight "trying to make sure that we didn't miss anything," particularly in Brampton, ON, where the Canadarm2 was being readied for launch, and at software laboratories in Vancouver. Press attention was intense as Hadfield flew between all the facilities in Canada and the United States, and put in long days training "in the pool" — the Neutral Buoyancy Laboratory where astronauts train in Houston.

"No Canadian had ever had this level of responsibility before on a space flight: [a] fully integrated and experienced mission specialist doing spacewalks," he said. "I'd already been Canada's first mission specialist on my first flight, but now here I am in the field first operating Canadarm, but now here I am building Canadarm2 and doing spacewalks. Obviously, a big deal; [today] it's on the five-dollar bill. It was a big deal. I took it very seriously and trained for it."

Astronaut Dave Williams echoed that sentiment after reading an early draft of my book. "Many feel the $5 bill has the two icons of Canadian space exploration — the Canadarm and a Canadian astronaut, representing an icon of technology and those Canadians lucky enough to have their dream of flying in space come true," he said.

After Garneau left the CSA in November 2005 to run for the Liberal Party, the agency went through a series of short-term presidents for the next three years. In quick succession, long-time CSA employees Virendra Jha, Carole Lacombe and Guy Bujold, as well as former Telesat employee Larry Boisvert, took acting or short-term roles as president.

Since none of these people were available for interviews for this book, it's difficult to speculate on what was going on internally at the CSA during this time. However, with so many presidents so quickly, some in the public worried that Canada's ability to set space policy may be diminished. As mentioned earlier, the Harper government was not always portrayed as science-friendly, and perhaps this had something to do with it, too.

There were also few Canadian flights during this period. Astronauts in this country had a five-year gap from going to space after Hadfield's flight, in large part due to the Columbia accident in 2003 suspending any opportunity to use the space shuttle in space for 2.5 years. But once the backlog began to clear, Canadians got their turn — MacLean and Williams each participated in spacewalks during construction missions, dubbed STS-115 and STS-118, in 2006 and 2007. And a record three Canadian astronauts flew in space in 2009: Payette on a shuttle mission, Thirsk on a long-duration space station mission (Canada's first) and space tourist and Cirque du Soleil co-founder Guy Laliberté, who paid the Russians for a brief visit to ISS. Thirsk, in fact, met both Canadians in space, although there wasn't much time for socializing with every-body's busy scientific experiment schedule, he said.[9]

Williams's story is an interesting study of the complex journey of becoming an astronaut. For a self-described curious kid from Saskatchewan, the 1960s were the golden era of exploring space and the undersea world. In 1961, Alan Shepard's historic first American space flight on the Freedom 7 Mercury mission sparked in him the dream of becoming an astronaut and exploring space. Attending an air show in St-Hubert, Quebec around that time, Williams watched Fern Villeneuve lead an eight F-86 aircraft formation through an incredible performance. So memorable, in fact, that Williams kept the Golden Hawks brochure and flew it in space on the Neurolab mission.

"Those pilots and the early astronauts inspired me to pursue my dreams. To me, it was the best way to thank them," Williams recalled. While flying in space was his ultimate goal, Williams was told that it was impossible as Canada did not have an astronaut program at the time. So at the tender age of 12 — in 1967, two years before the moon landings took place — he followed every kid's Jacques Cousteau dreams of the day and began pursuing his scuba training certification. This led to a lifelong obsession with human physiology and performance in extreme environments that brought him to medical doctor training and eventually back to human spaceflight — a big circle. In the end, he became an astronaut before becoming an aquanaut, when he spent a week on the Aquarius undersea research habitat in 2001, becoming the first Canadian to have lived and worked in space and the undersea frontiers.

"When I was 12 years old, I was [already] learning about . . . partial pressures, Boyle's law, decompression sickness, cardiovascular physiology, respiratory physiology and a whole host of other things that had human body functions under water," Williams recalled.[10] "That got me really interested in physiology and medicine. Of course, my mother was a former operating-room nurse as well."

Williams missed the first astronaut selection of 1983–84 completely because he was in final-year medical school at the time and busy studying and writing final exams. But seeing the application after the fact "reawoke the dream that there might be a possibility of becoming an astronaut in Canada," he said

As it turned out, his medical experience was a big asset in 1992, because by then the space agency wanted to get more experience in long-duration spaceflight and further understanding of the inherent issues related to decreased muscle density, loss of bone mass, fluid shifts, psychological loneliness and all the rest. In the early days of Williams's tenure, around 1993, he and Bob Thirsk were part of a Canadian proposal to send its own astronauts for long-duration missions on the Mir space station — an idea it eventually had to abandon due to affordability.

The interim solution was a one-week-long mission called by an obscure acronym, CAPSULS, and held at Defence Research and Development Canada in Toronto (which was then referred to by a different name, the Defence and Civil Institute of Environmental Medicine). Thirsk commanded the mission — the Canadians were initially hopeful he might have a chance to go on Mir someday — while Payette, Williams and McKay all served as crew members. "The timeline we followed was identical to one they were following on board a space shuttle research mission," Williams said. "We were actually eating space food and busy with experiments from international space researchers. It was a complete simulation. I think from my perspective, what was really exciting about that was our beginning to delve into true operational activity within the space medicine program at the CSA."

Williams built up his space medicine experience over the years, initially creating the space medicine program within the Canadian astronaut office, then with his Neurolab (neuroscience) focused space shuttle mission in 1998, but also as director of the Space and Life Sciences Directorate at NASA's Johnson Space Center in Houston. Williams managed a large team of civil servants and contractors with a budget of around US$275 million. This was only possible due to the dual approvals of NASA administrator Dan Goldin and the CSA's outgoing president at the time, Mac Evans, Williams said. The request came from influential NASA space station supporter and director of Johnson Space Center George Abbey, who wanted him for the job after seeing him work on Neurolab. Abbey was one of the chief champions for the final design of the International Space Station, who believed that international collaboration would be critical to the success of the program.

"His vision and commitment underscored the fact that the story of the space station is a story of collaboration. Here I am, a Canadian government civil servant [and] astronaut working as the deputy associate administrator of the office of space-flight," Williams argued. The first two co-chairs of the ISS multinational medical policy board were in fact not Americans, even though NASA is a majority-stake partner in the station. One of the chairs was Anatoli Grigoriev, whose many Russian medical honours include helming the Institute of Biomedical Problems that closely studies long-duration spaceflight. The other chair was Williams — representing NASA, "even though I was Canadian."

Only a year after the Neurolab mission, this Canadian's experience got the attention of NASA headquarters in Washington, DC, and he was asked to be the first deputy associate administrator for crew health and safety in the office of space flight at NASA headquarters. This put him among the top echelons of NASA management, just a couple of rungs down from administrator.

Williams held these senior NASA leadership positions for four years, a marvellous opportunity for a Canadian civil servant. Williams also found time to get more direct medicine experience in an isolated environment in 2001; he was one of the first astronauts to try out the underwater Aquarius habitat to simulate a space mission, kicking off NASA's NEEMO series that continues to this day. Williams never got his long-duration space mission, but he did get to do three spacewalks in a single shuttle mission in 2007 while managing a series of life science experiments on the space station.

STS-118 was the fifth mission after the shuttle's return to flight mission in 2005 and the first flight of Endeavour since 2002. As an ISS assembly flight, the crew would be installing the S5 truss segment, involving the famous Canadian hand-shake in space with the shuttle arm (Canadarm) handing S5 to the Canadarm2 on the space station. Williams was lead space-walker on two of his three spacewalks and set a new Canadian spacewalking record. "I hope that one of the current Canadian astronauts will get a chance to break the record," he said, sharing his dream for seeing Canadian astronauts continue to expand their contributions in space.

Williams almost got a chance at a fourth spacewalk, when it was discovered that the tile on the undersurface of Endeavour had been damaged during liftoff and mission control, meaning a tile repair might be necessary. Williams and crewmate Rick Mastracchio were initially slated to do the repair, which was cancelled after testing by NASA engineers found that the damage should not be a factor for re-entry. During and after the mission, the media focused heavily on the tile damage and penetration of the starboard radiator from orbital debris. These incidents clearly reinforced that human spaceflight can be fraught with risk.

While one astronaut was making moves in Washington, another was climbing his way to a high-flying rank in CSA central. In 2008, Steve MacLean — then retired from spaceflight — became president of the CSA and helmed it for five years. MacLean turned down the chance to comment for this book, so any insight about his term must come from outside observers.

His time at the agency was lengthy and a welcome break from the series of short-term presidents that had come before. With somebody more permanent in place, the CSA could perhaps finally work on a long-term space plan.

MacLean never succeeded, although the rumour was he put forward a proposal to the Harper government that did not meet what the government wanted.

There was documented progress under MacLean's tenure, though. Shortly after his arrival, the CSA announced that it would recruit new astronauts — the first such process in 16 years — and by May 2009, Jeremy Hansen and Saint-Jacques were ready to assume their basic training in Houston.

The timing was urgent. The other Canadian astronauts were now approaching their 50s and 60s. Canadians — and to be fair, many Americans as well — began to retire from the corps very quickly, since people beyond their 50s generally don't fly into space. Payette, Thirsk, Williams, and Tryggvason all left in a few short years; MacLean stayed on as president but retired from flying.

The urgency also arose from the fact that flight opportunities would decrease for the near future with the space shuttle program retiring. Crews would be reduced to three people a flight instead of seven, and astronauts would park on the ISS for six months instead of heading up and down again in a few weeks. Astronauts would have to be trained more heavily for generic tasks rather than specific ones, since it's hard to know what you will be required to do during a six-month period. Also, astronauts needed to be young and able to wait around a few years before heading into space, since there were fewer seats to go around.

Observers say that MacLean worked hard during these years to create stability for the CSA and to bring industry into discussions for a long-term space plan but that he was unable to get much done because the Harper government was not interested.[11] The Harper government, in a small gesture, did commission a review of the aerospace industry, chaired by former cabinet minister David Emerson. Perhaps they were hoping that the space industry would be placated by the committee's findings, but rather, in 2012, Emerson's group found that a lack of consistent funding to the CSA was crippling its ability to deliver programs.

Canada was about to send an astronaut into space — Hadfield, whose 2012–13 mission included being the first Canadian to command any spaceship, not just the ISS. His mission is well covered in many sources, but some of the remarkable things he achieved included managing a crew that dealt with an ammonia leak days before landing and helming the most scientifically productive mission to that date. These were excellent metrics of NASA effectiveness, and he did this while charming the world on social media in his spare time (with the support of extensive ground teams and his family).

After Hadfield, however, it was unclear when the next Canadian would go. It would be fully unfair to blame the Harper government for this situation because flight opportunities in general were diminished and the two new astronauts (Saint-Jacques and Hansen) had just finished their basic, 2.5-year training to qualify for flight opportunities. Canada's contribution is calculated to be 2.3 percent of the entire space station, and at the current pace of four flights a year, three people per flight, this naturally meant Canadians couldn't fly in space much more often than every five to six years.

What was more worrisome about the Emerson report was it pointed to a lack of vision and coherence for the space industry. Space projects take years or decades to plan, and Emerson's group pointed to some trouble spots, such as a lack of opportunity for commercial companies to participate — this in the era when start-up SpaceX was already launching rockets and making cargo vehicles, for example.

Emerson was a natural choice to helm the report because he had recently been the minister responsible for the CSA. His tenure was 18 months between 2004 and 2006 — not a very long time but long enough to begin to set some wheels rolling. And as the new Harper government took hold in those years, he noticed some issues with communications.

The first thing that struck him was the charisma of Garneau, a former astronaut who had been "something of a national celebrity" while leading the CSA between 2001 and 2005. "I would go on trips to planetariums and so on, which were of political interest to the government because of the audience," Emerson recalled.[12] "I used to marvel at how Garneau was a celebrity and I was just a mealy-mouthed politician that nobody cared about. I was very early on exposed to the romance and the charisma of the astronaut."

But when the lights went down and the negotiations needed to open up, Emerson saw an ongoing tension between the CSA and the industry department — as well as the ministry — about "who was telling whom what to do" and how to handle budgetary requests. "I got to see first-hand how something which, in my view, ought to be a fundamental strategic element of a national economic and industrial strategy, actually when it came down to the operations of government, was just another little section in the department fighting for increments of funding, or funding to not lose increments of funding. It struck me at the time what a backward way it was to be dealing with something as strategic as space."

He remembers budgetary decisions in particular being difficult: "Primarily bureaucrats fighting over basically the budgetary crumbs that the finance and treasury board put out there for people to fight over." He said this was partly a function of the environment, since funding was allowance-based and doled out by fiscal year, making it difficult to do long-term planning and research, which were "simply not part of the discussion."

"It always concerned me," he added, "that we didn't have a longer-term fiscal and economic commitment to space. It always felt that with Canada being somewhere up in the top five in the space world back at that time, it was a real opportunity to invest and build in terms of playing to our strengths and playing to a technology and an industry that was going to be absolutely crucial to Canada across a number of dimensions, whether it was coastal surveillance, security, communications, agriculture, climate management and so on."

Emerson moved out of the industry portfolio in 2006 as the CSA began its string of short-term presidents. He retired from politics and was contacted about 2011 — midway through MacLean's five-year term — to do the Emerson report because he had been supportive of the aerospace industry. By then, the Aerospace Industries Association of Canada wanted a complete review done of the industry; they requested Emerson do it through the usual bureaucratic and political channels. "They knew I had a strong sympathy with aerospace as a fundamental part of a Canadian industrial strategy," Emerson said, and he was duly appointed. He conducted a year of consultations across the country, touching on different parts of the aerospace sector, before his team wrote the report.

The AIAC's and the Canadian government's decision to do a separate report on space — outside of the aerospace sector — was something that Emerson did not anticipate whatsoever. But regarding space, he said, "You really got a sense as to how [it] was going to underpin so much of the Canadian economy. It was kind of an awakening for me. I remember thinking what a sweet spot space was, in terms of the Canadian economy, because it's a high-end, human-capital-intensive, research-intensive manufacturing sector, but it also, in a really clear way, underpins so much of the other aspects of the Canadian economy and social life, whether it's provision of medical care across different parts of a huge, expansive country, precision farming, patrolling the Arctic. We started to think about what a mess the whole Arctic patrol business was in Canada, with confusing organizational arrangements involving the Coast Guard, and the Armed Forces, and the RCMP, and so on, and how poorly we were really doing at it, in terms of providing the necessary surveillance and equipment."

Space is also an export-driven business in Canada, particularly to the United States. At the time, Emerson's group became very concerned with International Traffic in Arms Regulations (ITAR) that strictly regulated "dual-use technology," or technology that may have military applications beyond its original application, such as space-based operation. A small Canadian switch used in an American satellite, for example, may have difficulties being exported to a country such as China.

"I was hugely concerned, the deeper I got into it, about how we were really being prevented by the US security and military establishment from developing our aerospace sector, and particularly the space part of it, by restrictions under ITARs and restrictions on the people that could work in the manufacturing facilities who were dealing with these technologies," Emerson said, adding that he saw it as "an indirect but very lethal problem the space sector has dealt with in Canada."

MacLean resigned his post in 2013, just months after Emerson put out his report, and was replaced by Walter Natynczyk, who was fresh off a position as chief of the defence staff of the Canadian Armed Forces from 2008 to 2012. Natynczyk has a degree in business administration, which he jokes was really "in rugby and football." But more seriously, he argued that his experience as a soldier and in capital procurement has helped him understand all of Canada's armed forces and to put them "together into a coherent policy and into a coherent investment plan in order to achieve operational output."

"You can build exactly the same kind of case with a space agency, and that was the mission," said Natynczyk in an interview during his tenure as deputy minister of veterans affairs.[13] "So going into the portfolio, I spent time with the previous presidents of the space agency, those who were around and able to share their experience. And so I was able to understand that, over time, the success of the Canada space program really was linked back to individual political leadership in coordination with the president of the space agency . . . but it went back to having significant political support in order to endorse a long-term finance and investment plan."

He acknowledges that coming into his position as CSA president in the summer of 2013, he was faced with figuring out a direction for an agency that had not received an approved Canadian space policy since 1997. Further, Natynczyk had to work in an environment of "austerity," as he puts it — as the Canadian government was cutting back on projects out of concerns about increasing its deficit. The year 2013 was four years after the worst of the 2008–09 economic crisis, but the economy was growing slowly and seemed vulnerable to shock.

Canada's space sector was facing questions from its international partners as to whether it could participate in large astronomy projects, such as the James Webb Space Telescope and the OSIRIS-REx mission to visit an asteroid. The famous Radarsat Earth observation satellites were aging — in fact, Radarsat 1 died a natural death of old age in 2013, while Radarsat 2's technology was becoming outdated. Natynczyk's compromise was to craft a cost-neutral space policy framework. It didn't have the same weight as a government-backed long-term space plan with increased resources, but he said the important thing was it allowed the CSA to map its available budget with the priority projects.

In the years afterwards, Natynczyk's approach did bear some fruit, in that Canada solidified its commitment to all three of these pressing projects. Natynczczyk pointed out he was working amid an uncertain international space environment, since NASA administrator Bolden and his agency were in the middle of reimagining Bush's moon commitment as a set of more flexible destinations. It appeared that Canada was biding its time to see what the US would do, so it would be ready to swoop in. Natynczyk left the CSA in 2014 to take on a deputy ministerial portfolio, or at least, that is what Natynczyk and the Harper government said of the move. CSA appointed president Sylvain Laporte, who began work in 2015. As of early 2020, Laporte is on the road to retirement and a search for a successor president is underway.

So there was some progress during these years. Emerson argued in February 2019 — shortly before the big Canadarm3 and space plan announcements by the Trudeau government — that his own report had done little for the Canadian space sector since its release in 2012, an ironic statement given his urging for long-term thinking. And perhaps he spoke too soon. One of his recommendations — to have an independent board of experts helping the CSA with its funding decisions — was put in place, with Evans at the helm. This board has been busy running consultations across the country and acting as a liaison with Canadian industry, both factors that helped in the formulation of the long-term space strategy that was finally released in 2019.

Happily, even while the debate carried on in Canada about how to implement a space policy and whether it was a government priority, Canadian robotics carried on in space. Dextre's operations gradually became fully Canadian-based, and Canadarm2 proved even more essential in space station operations than ever expected.

Perhaps Canadarm2's most famous moment came when Parazynski — yes, part of the team that installed it — put it to use to stitch together a broken space station on his last of five space shuttle flights in October 2007. To set the scene: space shuttle Discovery brought up a set of solar arrays for the space station. These are vitally crucial for everything the space station does, since they generate power. Without power, there's no oxygen, no lights, no water, no reason for astronauts to stay on board.

Partway through the mission, after the folded-up arrays had been installed, the astronaut crew gathered to watch them be gently unfurled. Then trouble ensued: a frayed guide wire or something similar created two small tears in one of the arrays. If the astronauts and ground advisors couldn't fix this quickly, space station assembly would be once again delayed. NASA was eyeing retirement of the space shuttle at this point, in large part due to the fatal Columbia accident of 2003, so pushing the aging vehicle fleet into more years of service was another consideration on everyone's minds when delays occurred.

Luckily, on board was one of NASA's most experienced spacewalkers — Parazynski himself — who flawlessly executed the instructions from the ground. Equipped with a Mylar-tape "hockey stick," he rode an extension boom on the end of Canadarm2. And it should be mentioned here — because teams are never mentioned enough — that Parazynski didn't rocket up there by himself. The robotic arm was guided by Stephanie Wilson and Dan Tani, and watching Parazynski's progress outside was fellow spacewalker Doug Wheelock — with extensive interaction with Mission Control Houston.

Parazynski carefully threaded the solar array together using ad hoc "cufflinks" (wire and tape) fashioned in situ on the space shuttle in an "Apollo 13 sort of initiative" constructed by George Zamka and Peggy Whitson, he said.[14] The two astronauts transformed the Harmony connecting module into an impromptu metal shop, cutting wires three to five feet each in length and including pieces of aluminum shim stock cut into ovals, precisely measured with instructions from the ground.

Fashioning the cufflinks was no easy work for Parazynski's crewmates, and neither was his job of installing them on the solar panel. He was on the end of an extender on the end of an arm working right next to a fully charged electrical source. If he touched the wrong part of his spacesuit or tools to it, he could get a nasty electric shock — at best. "The wrist disconnection of my spacesuit even had to be specially insulated and taped so that no arcing would get into my spacesuit and ignite the 100 percent oxygen inside," Parazynski recalled in an interview.[15]

While Parazynski had dozens of hours of spacewalking experience and the ground crew trusted his work, the Canadian Space Agency's Ken Podwalski still remembers the tension while watching the astronaut run through the procedures that his team helped develop.

"You get into this remarkable complicated [situation] . . . everything piling together in terms of how bad a situation it is, but then look at the eloquence of the solution," he said.[16] Those elements were so simple that even a teenager could grasp them: Canadarm2, a mobile extender, cufflinks, and a hard-working team. "The way that all worked cooperatively together, it's just almost a fascinating . . . juxtaposition. I hate to use a fancy word like that, but it really kind of reflects one off the other," Podwalski said.

And it worked. Did it ever work. Not only did the team save the solar array, but incredibly, this fix — thrown together over 72 gruelling hours in Montreal and Houston and other space station centre locations — was still holding together beautifully nine years later, according to a photo posted on Twitter that Parazynski commented on. "Our repairs are still under warranty," joked the retired astronaut in 2016.[17]

Obviously, the space station assembly story is about far more than Canadarm2, but do understand this — it was remarkable that modules manufactured in Russia and Europe and the United States, for the most part, came together beautifully in orbit. These modules weren't made together and had never mated together or even been seen together until they joined in space. Research modules, a robotic arm, a huge window and much more came up, piece by piece, on the space shuttle. Over more than 15 years, the space station grew and grew and grew and in fact, still grows — but at a slower pace — today.

More recent ISS additions after the space shuttle program ended include an inflatable room — the Bigelow Expandable Activity Module — that could pave the way for more light-weight spacecraft in the future, and docking adapters that will allow commercial spacecraft to dock with the ISS in the coming years. (In March 2020, amid the novel coronavirus pandemic, Bigelow laid off its employees, according to media reports; the effects on its programs remain to be seen.) More recently, NASA opened up the possibility of allowing commercial companies to dock temporary modules with the space station and fly their own private astronauts up to do experiments in them. That's not realized yet and still very much in the works, but the ISS continues to change and evolve even after more than 20 years of operations in space.

CHAPTER 6

More than just visitors

> Boromir: "But of that perilous land we have heard in
> Gondor, and it is said that few come out who once got
> in; and of that few, none have escaped unscathed."
> Aragon: "Say not unscathed, but if you say
> unchanged, then maybe you will speak the truth."
>
> — J.R.R. Tolkien, *The Lord of the Rings*

Before I go any further, I need to talk to you about how weird
the International Space Station environment is. I don't just
mean the microgravity and the fact that you're living in a spaceship. I also mean that Americans, Russians and other citizens
can flow seamlessly from module to module without needing
to worry about border control.

The ISS works a little bit like the eurozone, in that each person
on the station has rights to move into other countries' territories. There are no border guards monitoring each space station
doorway (which is great because you'd need to fly these people
up, feed them and keep them occupied and alive for months at a
time). Instead, the ISS is under an intergovernmental agreement
(IGA) signed by the 15 or so space station partners.

"Because we are cooperating internationally — we can essentially work out a way that we can go into each other's modules whenever necessary," said Michael Dodge, an assistant professor at the University of North Dakota who has a specialty in space law from his previous studies at McGill University in Montreal.[1]

There are a bunch of ways this cooperation can work, he explained. The IGA specifically sketches out the role of the crew commander (which was Hadfield in 2013) and how that person is in charge of the way things work. The usual export controls between countries are also not an issue, he said — which is a good thing because every astronaut should know how to operate the other modules in case of emergency. (Export controls may apply in small measure to individual experiments, especially those with commercial potential, he added.)

But where the IGA gets most interesting is in its discussion of criminal law.

"It essentially creates an extradition treaty between the United States and the Russian Federation that does not exist on Earth," Dodge explained. "In the instance where there might be a criminal activity — if the state whose citizen has conducted that . . . alleged criminal activity refuses to prosecute or [do] something about it, then the other state has a right to prosecute that person. Since the United States and the Russian Federation are the two big partners there, that could mean potentially extradition in a way that does not exist on Earth, which is kind of interesting to see."

Space has always been a little more open-source than Earth. If you look back to the Outer Space Treaty formulated in 1967 — when the Cold War was still very much a thing — it includes stipulations such as banning nuclear weapons and says that no nation can claim sovereignty over celestial bodies such as the moon (upon which NASA was about to land humans). There's also a Rescue Agreement — which flows from certain articles of the treaty — that says if an astronaut accidentally lands in a foreign country, they are supposed to be taken care of and returned safely to their home.

So is the ISS utopia? Does everyone there always get along, regardless of nationality and creed? It seems unlikely, but NASA and the other agencies keep any disputes quiet for obvious privacy reasons (and the possible international implications). And while the agreement has remained strong through many international spats, there was one recently that briefly looked like it was going to break the chain of missions.

That was the moment in 2014 when a Russian official quipped that if NASA wanted to get its astronauts up to the space station, it would need to use a trampoline. This Russian official was the deputy prime minister, Dmitry Rogozin, speaking after a series of US sanctions in the wake of a crisis in Ukraine.

The causes are complicated, but as a short description — in November 2013, then Ukrainian prime minister Viktor Yanukovych rejected a deal that would have created closer ties between Ukraine — which is located right next to Europe — and the European Union. Mass protests eventually made Yanukovych leave the country, and Russia's response was to annex Crimea and then move troops into eastern Ukraine as well.[2] The Americans were not happy about this and imposed economic sanctions.

So why a trampoline? Rogozin was pointing out that the Russians, and only the Russians, could send people up to the space station for the time being. The Americans had retired their space shuttle in 2011, and until new commercial crew vehicles were ready, they had an agreement to use the Russian Soyuz for astronaut transportation — at a cost of tens of millions of dollars per seat. NASA carefully played down the Rogozin comment, and launches continued on time and without fail. But Congress wasn't happy, and neither was United Launch Alliance (ULA) — the US space launch company who relied on a Russian RD-180 engine for its mighty Atlas V rocket. ULA will be moving away from the RD-180 for future rocket types, and the international dispute eventually simmered down in the public.

It is remarkable that the ISS continues to work as well as it does. To their credit: NASA and Roscosmos continued relations through the situation, and no launch was delayed for geopolitical concerns.

But sometimes changes on the ground do prompt even subtle changes in space. Russian journalist Anatoly Zak, who is nearly unique in the space reporting world due to his Russian heritage and long history reporting in English, pointed out that a wall on Russian module Zvezda had photos of Korolev (who founded the Soviet space program) and Gagarin (the first person in space). But that changed in 2014 shortly after the crisis occurred, Zak said.

"The Russian Orthodox Church, which has long had a broad influence on the country, including spacefaring activities, quickly spoke out as one of the most ardent supporters of the ultra-conservative and nationalist policies adopted by the Kremlin," he wrote. "In 2014, Crimean-born cosmonaut Anton Shkaplerov, now a symbol of the Russian space program, made a widely advertised visit to his homeland before his flight to the station. Upon his arrival at the station in November, photos of Shkaplerov and his crewmates showed a backdrop of Orthodox icons and crucifixes — Korolev and Gagarin had been moved quite literally out of the picture."[3]

But it would be unfair to only talk about the frustrations of the partnership because, overall, relations appear to be smooth between astronauts and cosmonauts. Canadian astronaut Bob Thirsk, who flew the first Canadian long-duration mission on the ISS, said that a typical Canadian walking around in Moscow would feel a cultural difference. But among astronauts and cosmonauts — a highly operational and scientific group of people with a strong spirit of adventure — that barrier disappears because there are similar experiences and mindsets.[4]

John Chapman (left) was a Canadian space researcher who saw satellites as key to communications and monitoring of Canada's vast space. Although he died in 1979, he greatly influenced the formation of the astronaut program and the beginning of the Canadian Space Agency in the years afterwards. Chapman is seen here toasting the success of Canada's first satellite, Alouette, with fellow Canadian space researcher LeRoy Nelms.

Canada's Avro Arrow program would have seen an advanced fighter jet enter service in the 1960s. When the John Diefenbaker administration cancelled the Arrow in 1959, dozens of Canadian engineers instead joined NASA's young space program to get ready for the Apollo moon missions late in the 1960s.

Material republished with the express permission of: *Ottawa Citizen*, a division of Postmedia Network Inc.

NASA tested space shuttle landings using a test vehicle called Enterprise, named after the famous *Star Trek* ship. In 1983, Enterprise visited Ottawa's airport as the Canadian government prepared to launch its astronaut program.

NASA

NASA astronauts Dick Truly and Joe Engle, flying space shuttle mission STS-2 in November 1981, expressed pleasure at the performance of Canadarm on its first flight. The robotic manipulator system became crucial to astronaut space work, performing tasks such as retrieving satellites and helping to build the International Space Station. Subsequent robotics such as Canadarm2 and Dextre assist today with ISS spacewalks, and a Canadarm3 is planned for NASA's Gateway space station at the moon in the 2020s.

Astronaut Marc Garneau flew in space in October 1984, only a year after he was selected, making him the first Canadian in space. The Navy-trained Garneau, careful to not overuse communications in space as a rookie astronaut, was dubbed "The Right Stiff" by some Canadian media. He flew twice more in space before helming the Canadian Space Agency. As of mid-2020, the long-time politician is Canada's Transport Minister in Canadian prime minister Justin Trudeau's Liberal government.

Roberta Bondar was the first Canadian woman in space, having served previously on the Canadian Life Sciences Subcommittee for the Space Station. She was designated the prime payload specialist for the first International Microgravity Laboratory Mission, a coup for Canada's young space program that generated a lot of publicity. Shortly after her January 1992 flight, Bondar was told to resign. The reasons remain unclear a generation later. Bondar today runs a wildlife foundation, is a professional photographer, and remains active in space outreach.

Steve MacLean, shown here honouring the gymnastics team he used to be a part of, flew twice in space. His achievements for Canada's space program include testing early versions of the "vision" system that guides robotics in the challenging lighting conditions of space, performing spacewalks and helming the Canadian Space Agency. MacLean continues working today in the private sector.

Canadian astronaut Bjarni Tryggvason (right) is shown here during training for mission STS-85, along with crew member and NASA astronaut Stephen Robinson. Tryggvason's engineering work in space includes serving as project engineer for the Space Vision System Target Spacecraft and developing the Microgravity Isolation Mount that operated on the space station Mir (then a Russian vehicle) between 1996 and 1998. Tryggvason continues working today in the private sector.

Julie Payette (centre) talks with a ground worker during the lead-up to mission STS-96. The astronaut said she was impressed with the workers' commitment to ensuring safety in space. Payette's numerous achievements for Canada include flying twice in space, operating the Canadarm, doing research in artificial intelligence and taking on a fellowship at the Woodrow Wilson Center for International Scholars. She later became Governor General of Canada, a position she still holds in mid-2020.

Chris Hadfield gives the thumbs-up during his first spacewalk in 2001. Hadfield flew three times in space, visiting both the Mir space station and the International Space Station. He was the first Canadian to command the International Space Station in 2013; the second was NASA astronaut Drew Feustel (a dual Canadian citizen) in 2018. He also held several management positions. Hadfield remains a frequent commentator on space today for numerous organizations and as a public speaker.

Canadian astronaut and medical doctor Dave Williams takes part in the NASA Extreme Environment Mission Operations (NEEMO) 9 underwater excursion, off the coast of Key Largo, Florida. Astronauts regularly do simulations in different isolated environments to prepare for spaceflight. Williams flew twice in space, performed spacewalks, participated in the Neurolab Spacelab Mission and was director of the Space Life Sciences Directorate at the NASA Johnson Space Center, as well as many other achievements. Williams today runs a business and does speaking engagements, among other activities.

Canadian astronaut Jenni Sidey-Gibbons (left) talks about how fellow astronaut Roberta Bondar (right) inspired her in a Canadian Space Agency video. Sidey-Gibbons is a fire researcher who joined the Canadian Space Agency in 2017, graduating from basic training in early 2020.

Canadian astronauts Julie Payette (left) and Bob Thirsk (right) met up in space during Payette's shuttle mission, STS-127, in July 2009. Medical doctor Thirsk flew twice in space and was the second Canadian to have a long-duration mission on the International Space Station, after NASA astronaut and dual citizen Gregory Chamitoff in 2008. Thirsk also met in space with Canadian entrepreneur Guy Laliberté during Laliberté's tourism flight to space in 2009. After resigning from CSA in 2012, Thirsk became president of the Canadian Institutes of Health Research before moving into a private career of space outreach and public engagement.

Joshua Kutryk made the finalist round of astronaut selection in 2009 and was not ultimately selected, but the Canadian test pilot returned and gave it his all in 2017, as seen during a swimming test here. Kutryk was selected for the 2017 astronaut class, whose training was overseen by Canadian astronaut Jeremy Hansen. Kutryk graduated from basic training in early 2020.

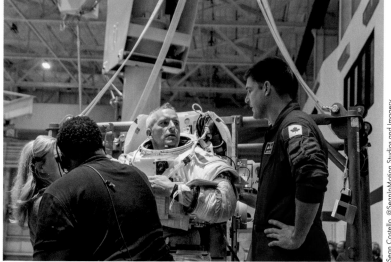

Sean Costello, @SeanInMotion Studios and Imagery

Canadian astronauts David Saint-Jacques (left) and Jeremy Hansen (right) confer during spacewalk training at NASA's Neutral Buoyancy Laboratory in Houston, a giant pool that includes a scale model of part of the International Space Station. Medical doctor Saint-Jacques waited nine years after his selection in 2009 before flying in space, while fighter pilot Hansen (also selected in 2009) is still waiting for a spaceflight assignment as of mid-2020.

Sean Costello, @SeanInMotion Studios and Imagery

David Saint-Jacques (centre) and two crewmates climb aboard the Russian Soyuz rocket that will carry them into space from Baikonur, Kazakhstan, on December 3, 2018. This was the first human spaceflight attempt since a crew rocket abort barely two months before, and this time it was successful, demonstrating the value of international teamwork in safely keeping the space program going. Saint-Jacques performed a spacewalk and helped welcome the first uncrewed SpaceX Crew Dragon spacecraft to the International Space Station, among other achievements.

And when it comes to politics? That is usually left aside, he said, because the group has other goals. "In spite of our cultural, historical and ideological differences, my crew rarely talked of them. We're very like-minded. Advancing the capabilities of humanity in space through collaboration and innovation was a more important concern for us."

What the crews did instead in space — at least when Thirsk was up there — was to gather at meals and to have cultural exchanges. "I would talk about how the Canadian parliamentary system works and describe question period. I'd recount stories of our Indigenous peoples and early explorers. I'd describe how hockey has become a part of our national identity. I'd explain how our participation in the shuttle and space programs has bolstered Canada's foundation in exploration and diplomacy."

Then it would be Thirsk's turn to listen. "My crewmates shared the histories of their countries. Russia has apparently been invaded eight times over the last 200 years. Perhaps these actions have resulted in a heightened territorial mindset for Russian leaders and might've factored into political actions like the more recent annexation of Crimea. But the point is that mealtimes aboard the station would be opportunities for my crew to discuss a variety of global topics — not just scientific and technical ones."

He said it was difficult to put his finger on why the crews tended to work well together despite international differences, but the technical expertise, willingness to ignore political and programmatic squabbles on the ground and these cultural exchanges all seemed to be elements. "On the ground, like I said, there are cultural differences between members of societies. But on orbit, we seemed to be one."

All astronauts get cultural training. Part of a typical astronaut's preparation for the ISS includes engaging in homestays with families in Moscow — in between their training in Star City — so that the astronauts can become more fluent. Russian fluency is necessary because the astronauts may be called upon, in space, to operate a spacecraft system or Russian module at a moment's notice. They will also need to understand the technical instructions from Mission Control in Moscow.

"We fly the Soyuz as the prime flight engineer, which requires years of technical systems and orbital mechanics study in Star City, and then operating the spaceship through launch, orbit and landing, solely in Russian," pointed out astronaut Chris Hadfield after reading an early draft of this book.

Technical expertise is supplemented with experience in cultural relations.

"On two occasions, I lived with Moscow families," Thirsk said. "One occasion for six weeks, and another occasion for three weeks. I was embedded in Russian family life during weekday mornings, weekday evenings and weekends. During the workday, I was in language training. So it was total immersion — similar to French immersion programs in Canada. It was a means to not only pick up the language, but also to appreciate the wonderful arts, entertainment, culture and family life of Russia. A rare and privileged opportunity to get to know people."

Thirsk spent his time off in the evenings getting to know Moscow — looking at its Impressionist collection at the Pushkin State Museum of Fine Arts, attending the ballet or the symphony, going to hockey games. He also exercised outside, allowing that might be easier for a Canadian than say, a Californian. "Some people would say, 'It's too cold to go jogging tomorrow.' Well, no, it's not. I've been jogging like this all the time."

This all helped Thirsk build his cultural and linguistic fluency. "Fluency" is of course a vague term and varies with astronaut experience, but even the most junior astronauts can follow commands from their Russian space station superiors and understand the essentials of technical conversations. While at a press conference in Baikonur, for example, I noticed that Saint-Jacques was capable of answering questions from the press in Russian. (In particular, I remember one discussion in Russian where he was talking about how his experience in the North of Canada helped him become used to isolated environments.) The more junior McClain preferred to stick to English in her responses, but she had passed the Russian-only oral exams weeks before to get qualified for the Soyuz spacecraft, which indicates a high degree of technical confidence.

"They're stand-up oral exams," Thirsk explained. "After studying a spacecraft technical system for two or three weeks, it is time to assess our understanding of and operational competence with the system. We will arrive on the morning of the exam to a room filled with system experts — experts from the Star City training centre but also from Energia and other industrial contractors. The experts are seated and we stand alone facing them at the front of the room.

"The experts fire questions at us for, if we're lucky, a half-hour," he continued. "But if you're having a bad day and there is any doubt that you don't know your material, the experts could ask questions for an hour. At the end of the exam, we receive a score from one to five. One means you don't know your material; five means that you are completely proficient. The first exam was a bit stressful, but we quickly adapted to this Russian way of doing things."

And knowing Russian also blends right back into those dinners in space, Thirsk said. He would make every effort to speak Russian to his Russian crewmates, and if words began to fail him, he would even try throwing "as many Russian words as possible" into a sentence. "They probably didn't even notice," he said, "but if I made no effort at all to speak their language, they probably would have noticed."

Several Canadians have spent tours of duty aboard this space station, so we should get a picture of what it looks like. NASA has a few YouTube tours of the orbiting complex, where astronauts patiently move through it room by room and show viewers what it's really like. The thing that always struck me was the narrow opening between the "Russian" segment (all Russian modules) and the "American" segment (which includes modules from Europeans and the Japanese). Astronauts need to squeeze through a hole barely wider than their bodies, and underneath are bags full of stuff in storage.

I asked Thirsk to talk to me about the character of the individual modules, because there's only so much you can see on the videos. He told me that nobody had ever asked that before, then said it's true that each room feels a little different than the others.

Let's start with the spot that Western viewers of space are most familiar with — the Japanese Experiment Module.[5] I say it's "familiar" because it's the nice open space where the astronauts do their press conferences and chats with children. So it's a frequent backdrop to those who have watched these space events, with its string of flags cheerfully adorning the background. It's also the spot where astronauts operate the small Kibo robotic arm and the "outdoor" Japanese facility, where they run exposure experiments or shoot off tiny CubeSats into space from time to time.

Thirsk used to sleep there, he told me. "For the first four months that I was on orbit, there were only five sleep stations for six crewmembers," he explained, which meant one crew member wouldn't have a spot to go. "And I volunteered to sleep out in the open space of the Japanese module. The second shuttle flight to visit the station during my increment delivered the sixth sleep station. This last sleep station was temporarily set up in the Japanese module before moving in later months to its permanent location in Node 2," Thirsk added.

Leaving Japanese territory, you suddenly slam into a major European intersection. Called Node 2 or "Harmony," it allows you to branch off into several directions, depending on your needs. If you fly on straight from the Japanese module, you'll end up in the European experiment area — Columbus. Turn left or fly "up" (assuming you're in the same standing position you would be on Earth), and you'll run into docking modules that may or may not have spacecraft attached. But if you want to see the rest of the space station, you'll need to make a right turn to head down the long, long axis towards the Russian segment.

Oddly, the first thing you'll come across as you go through Node 2 is four crew sleep stations — one radiating out in each cardinal direction. This means you could be sleeping upside down (like a bat) or standing up (like a horse), or on your side. But again, that assumes directions that make sense in gravity. In space, there's no up or down, and the astronauts are just happy to have a small private space. It's just big enough to use a sleeping bag and to check your emails on a small laptop computer attached to the wall so it doesn't float away.

The next stop, past the bunks, is the US Destiny Laboratory, where the Americans do most of their experiments. You may have heard that a certain percentage of experiments on the space station are done by commercial companies. Well, this is where they are located — along with a suite of experiments from governments (including Canada) and university partners. Hidden behind front panels of many US Laboratory racks, Thirsk added, are vital life-support systems for the space station — such as the oxygen generator and the carbon dioxide "scrubber" that takes this toxic gas out of the air as astronauts breathe it out. And if you fly "up," you'll encounter a port to one of the external stowage pallets, which host spare parts in space for space station repairs — meaning stuff like pumps or joints.

But continuing straight ahead, you'll run into Node 1 or "Unity" — the very first United States space station component. That's where the tiny space station kitchen is located, as well as a foldable table (which is more a convenient spot for conversations than to put your food, since everything floats). To your left is an airlock to do space station spacewalks — which, handily, includes another external stowage pallet for spare parts. Straight ahead is Russian territory, and to your right — where we'll move next — is Node 3 or "Tranquility."

Tranquility, even though its name evokes memories of NASA's Apollo 11 for readers of a certain age, is actually another European module. It's also a spot that people may recognize, because if you fly "down" while in the module, you'll end up in the Cupola. This part of the space station is so new that Thirsk did not get to enjoy its 360-degree view in 2009; it only flew to the space station in 2010, near the end of construction. Astronauts use it constantly for relaxation and also to do Earth observation activities.

A lot of publicity photos are done in this area because it's so pretty. Oddly, Node 3 is also host to a rack of exercise equipment and a toilet, so it's hard to say how tranquil the Cupola is — or how good it smells. But the view out of the Cupola is most definitely worth the trouble, as astronauts lucky enough to use it always say they enjoyed the experience.

To the left of Node 3 is an interesting area called the Bigelow Expandable Activity Module (BEAM), which could represent the future of how space stations are built. So far, we stuck to aluminum — a strong, lightweight metal — but BEAM is made up of durable inflatable material. It's there as part of a long-term test by Bigelow Aerospace to see how viable inflatables are for long-term space exploration — how well they can stand up to radiation and if they are as resistant to micrometeorite strikes as the space station hull is.

Then to the right is another room with a storied history: the European Permanent Multipurpose Module. It used to be a big storage canister brought up on shuttle flight after shuttle flight for awkwardly shaped equipment. But when the shuttle ceased flying in 2011, there was no need to keep hauling it back and forth to space. So the module transformed into another space station room. It's used today as part storage facility and part scientific laboratory.

Now it's time to head over to the Russian segment. Go back to Node 1, then make a right turn and over the hump that represents the pressurized mating adapter (PMA in astronaut parlance). Floating through the narrow hatch, you'll notice bags of stuff underneath. I saw multiple astronauts narrating tours of the space station, but the stuff went unexplained — so I asked Thirsk what's inside there.

"The stowage there is strategic," he explained. "We use PMA1 for stowage of items that we need to access frequently and easily — for example, pens, pads, hand towels, toothpaste, toothbrushes and razors. The white stowage bags for these items are stacked no more than one layer thick. It would otherwise be impossible to float through the narrow PMA passageway. In other station modules that are deliberately designed for stowage, such as the Japanese logistics module, we stack stowage bags three layers thick. It is more challenging to find items in those modules."

The first stop is the Functional Cargo Block, which is abbreviated to FGB (a shortened version of its Russian name). The Russians also require their own storage space, so passing through, a visitor will see lots of containers — especially urine containers, Thirsk recalled, since he remembers going in there constantly to change them out. Even the outside is a storage zone, as it includes large tanks for propellant and for water.

Flying straight ahead, we finally find ourselves at the opposite end of the space station to where we began. Now we are in the heart of the Russian zone, which is known as the Service Module. "The Service Module is the brain, the nerve centre, of the Russian segment of the station," said Thirsk, a description that is probably very meaningful to him given his medical training. The cosmonauts have sleep stations, computers used to monitor station systems, a ham radio, several windows with an assortment of cameras and lenses nearby, a rear hatch that usually leads to a Soyuz vehicle or docked cargo vehicle, a food pantry and a table for dining.

"It's where things are commanded from, [through] the computers, to interface with all of the systems and payloads on the Russian segment," Thirsk explained. "The Russian cosmonauts spent much of their working day in the Service Module. We joined them there for some of our meals, for Earth observation photography, to use the ham radio and simply to chat throughout the day."

And at the onset of many types of emergencies, the crew — even those who happen to be working at the opposite end of the space station — will stop everything and rapidly fly all the way back through the many rooms and that oh-so-narrow opening to muster in the Russian Service Module. From this central post, the commander and crew will respond to onboard contingencies like fires, depressurization and toxic atmospheres. If ammonia from the external cooling loop, for instance, were to leak into the cabin atmosphere, the crew would quickly don gas masks — stored in the nearby FGB — and take corrective steps to minimize the contamination. "And if there were to be a particularly intense solar flare, it is the Service Module where the crew would take shelter. This Russian module includes a lot of massive equipment that could somewhat shield astronauts from the incoming radiation."

Astronauts get a safety briefing within minutes of arriving on station, because there are so many ways in which things can go wrong. The worst nightmares for astronauts usually include fire — especially given the history of Mir, which faced a severe fire that came close to forcing an evacuation — and depressurization. So the space station is designed with safety measures for several scenarios. The astronauts train extensively on the ground so that their response to an emergency comes through muscle memory. There's no time to read checklists when you have seconds to spare.

Upon getting to their new home, all astronauts are retrained on how to close a hatch quickly. That's no small matter because, as every submarine disaster movie has ever shown, it's all too easy to inadvertently trap somebody on the other side. The crews also are shown where each flashlight is — usually stowed at hatch entrances. They are "humongous," Thirsk said, presumably so you'll find them even when fumbling in the dark.

"Also, next to each hatch, there's also cable cutters," he explained. The original design and the initial configuration of the station had all utilities (power, data, video, fluids) pass from module to module through bulkheads. With the addition of payloads and systems over the ensuing years, additional small cables had to be routed across hatchways — not a decision that was taken lightly. So if a hatch needs to be urgently closed and disconnection of the cable is not a speedy option, then cable cutters would be used by the crew to sever those data or video cables. And if you think changing the wiring in a house is complicated, imagine doing that in microgravity inside a spacesuit.

Since astronauts spend so much time training for emergencies, it's easy for media to focus on these processes and procedures. But the everyday living — that is also endlessly fascinating. While Hadfield was in space in 2012–13, he and the CSA team uploaded several videos showing him doing familiar-yet-strange routines such as cutting nails and brushing teeth in space. In microgravity, everything floats, and it's easy for small bits to get lost. Hadfield's tip was always to wait a day or two and then check out all the air filters, because sooner or later debris will get caught in there.

At times, an astronaut's life in orbit can be scheduled in five-minute increments. Astronauts live on London time (or more properly, Greenwich Mean Time) and only get their detailed daily schedules early in the morning. "While the crew sleeps, the ground support teams at mission control centres in Houston and Moscow are awake and busily putting together the work plan for the upcoming day. Their objective is to finalize, collectively review and then transmit the timeline up to the station before the crew wakes. The night shifts at mission control are busy shifts," Thirsk said.

Thirsk would rise at 5:30 a.m. or so and take a moment of self-care — grabbing a weightless coffee — and then review the schedule while sipping his pouched hot beverage with a straw. His next priority was commandeering a handful of Ziploc bags to collect the tools and supplies he would need to complete each of his assigned tasks.

This would take nearly two hours. At 7:15 a.m., the entire six-person crew would gather in the Russian Service Module and eat breakfast together. Mealtimes are almost the only time crew members get to see each other, because their lives are so packed with experiments otherwise. Breakfast needed to end around 7:30 a.m. for the Daily Planning Conference with the ground. This overview briefing of the day's activities takes about 15 to 20 minutes, Thirsk said, because the crew needs to chat with different mission controls as they fly around the world — NASA, Europe, Japan, Russia.

The space station crews tend to be divided along Russian or non-Russian lines during the day, since Roscosmos manages its own scientific programs separately. Thirsk observed that the non-Russian astronauts often worked for hours with little verbal communication with mission control. If a particular crew was comfortable with oversight by the ground via onboard video, they would turn on the video cameras throughout the station at the start of the workday to provide the flight control team with "situational awareness" (while some crews turn it off for privacy, Thirsk said it is handy in mission control — a good

CAPCOM knows not to bother an astronaut when they look very engaged in a task, unless it's urgent.)

The Russians, by contrast, preferred to chat with the ground as they did their tasks. Neither is right or wrong, just a cultural preference. And Thirsk noticed something else while floating by a Russian colleague. "If you eavesdrop on a conversation between a Russian cosmonaut and Moscow, it often seems to be protracted and agitated," he said. Thirsk was initially puzzled after his arrival on the station. He thought he was overhearing arguments between a Russian crewmate and mission control. His crewmate would deny this and reassure Bob that it was just a normal conversation with Moscow.

It was only after a few more of these kinds of exchanges that Thirsk began to understand. "Aha! they're not actually arguing," Thirsk explained. "It's just when they discuss important matters, their conversation gets especially animated and fast paced. It's normal Russian culture to wear their hearts on their sleeves."

It's hard to describe a typical workday for an astronaut, because the demands of the space station change. One astronaut might be assigned to work with a balky toilet for an hour. Another could be working on an isolated experiment in a single module, cut off from their crewmates, for several hours. There could be a call with schoolchildren in the middle of the day. While certain things have to take place at a fixed time — like a press conference — NASA does give the crew flexibility to move around their tasks as needed. After all, the astronaut has the best insight into onboard operations, Thirsk explained. "While I know what tasks are expected to get done before the end of the day, I am at liberty to reschedule the day's activities as I feel best. I have an understanding of limitations (such as availability of tools) or onboard schedule conflicts that the ground cannot possibly see. We appreciate the ground delegating this authority to us."

But one thing that rarely changes for astronauts is their exercise period — two hours a day, including set-up and take down of equipment. NASA aims to never overschedule an astronaut so that they feel tempted to eat into their exercise time, because that's counterproductive to their crew's long-term health. Even with so much workout time, a typical astronaut on a six-month mission will have trouble standing up, working and driving for at least a few weeks upon return.

Life in microgravity makes weightlifting useless, so NASA instead has an Advanced Resistive Exercise Device (ARED) developed by engineers at the Johnson Space Center. Astronauts push against a piston-and-cylinder assembly, which simulates weights and makes their muscles work hard. (A previous version used elastics, but the astronauts grew too strong over time for it to help completely.) For cardio work, there's a space station treadmill. A harness worn over the shoulders and around the waist while exercising holds the astronaut down. The straps of the harness can be tensioned to provide a gravity-like loading to the hips and legs while running.

While everything is tightly scheduled on weekdays, usually Saturdays and Sundays include time off for the crew. But — "because we're type A behaviour," as Thirsk pointed out — often the astronauts will do some extra tasks. This is when many of Hadfield's educational videos were produced, for example. Or astronauts might volunteer for additional science, or a little extra cleanup to make the week easier.

Otherwise, there's some time available for astronauts to catch up on news through their slow internet connections, to watch a TV show that NASA uploaded through the Tracking and Data Relay Satellite network, to do personal phone calls or to simply gaze out the window.

For Thirsk, this "off time" offered a valuable opportunity to reflect on his mission. Looking back, he never thought he would have the chance to spend six months in space. After his first flight in 1996, he came home and did the usual set of post-flight briefings and press conferences. Then he went to Mac Evans and asked what was next for his career. Evans wanted to put him in a management position, but Thirsk had minimal management experience. So the Canadian Space Agency sent Thirsk to the Massachusetts Institute of Technology for a year, where he completed a Master of Business Administration degree at MIT's Sloan School of Management.

Only weeks after returning home in 1998, Thirsk was just about to step into his new position when he got a second request. Canada had reformulated its contribution to the International Space Station and the powers that be offered Thirsk the opportunity to fly as Canada's first long-duration astronaut sometime in the next decade. He quietly began training but saw his flight date pushed back when the Columbia space shuttle broke up above Texas on Feb. 1, 2003, killing its seven crew members. Nobody flew a regular space shuttle flight until 2006, and as a Canadian, Thirsk was much further down the flight-priority list than an American. So he didn't make his journey into space until 2009.

While the extra time on the ground could have been frustrating, Thirsk said it allowed him to pick up more of those generic skills that are so valuable during a long-duration mission. Like any mechanic or scientist, an astronaut needs to know how to troubleshoot things on the fly. On a complicated machine, this could mean needing to know the guts of the electrical power or thermal control systems, Thirsk said, because the crew are the people who can function as the onboard eyes, ears and hands of the team. NASA has a handy feature called "just in time training," which sends up videos, schematics and other technical information the astronauts might need to repair specific systems, but it's good to at least be comfortable with expected maintenance and common repair jobs before leaving the ground. After all — the more time you save in maintenance in space, the more time you can spend on research. And being careless isn't an option. One hasty repair could ruin the space station experience for everyone.

Thirsk also used the years on the ground to train as Dave Williams's backup to be the commander of NASA Extreme Environment Mission Operations (NEEMO) Mission 7 in 2004. Williams stepped down from the role after he was diagnosed with cancer but ultimately regained operational status to be the commander of NEEMO 9 in 2006 and later fly in space on STS-118 in 2007. Thirsk said this mission was likely instrumental to helping him during his long-term stay in orbit. In situations like NEEMO, wilderness training or the underground work some astronauts do as part of CAVES (Cooperative Adventure for Valuing and Exercising human behaviour and performance Skills), astronauts must work in teams under arduous and sometimes stressful conditions. You learn a lot about yourself when you're squeezing through tight spaces, struggling to set up underwater equipment in an awkward scuba suit, or carrying heavy equipment in a –20 degree blizzard in the wilderness.

"As I get older, I'm wondering whether or not these soft skills — self-care, self-management, teamwork, followership, leadership, cross-cultural sensitivity — are the difference between a good astronaut and a great astronaut," Thirsk said. "Candidates are considered for astronaut recruitment for their impressive operational backgrounds and unique technical skills, and their ability to acquire the necessary knowledge to operate the station and the skills to perform robotics, EVA [spacewalks], assembly, maintenance. But not all candidates have the necessary soft skills to perform well as a team member during long-duration flight."

So far, this all sounds like a heroic adventure for Thirsk — living underwater, doing science while floating in orbit, talking Russian fluently after taking in the ballet and hockey games. But he found himself in a vulnerable position when he finally made it into space again. Thirsk noticed something troubling one day in orbit: he couldn't read one of the checklists while packing up a spacecraft to haul equipment away from the space station. It was in eight-point font, he recalled, and he struggled to read it even after holding a strong light to it.

At 55 years old, it would have been reasonable for Thirsk to have changes in his eyesight — after all, many of us are using bifocals or starting to worry about cataracts at that age. But this change appeared a lot more suddenly than what could be expected from just normal aging; he was reading these check-lists, no sweat and no strain, just a few weeks beforehand. So Thirsk mentioned it over dinner to fellow astronaut Mike Barratt, a 50-year-old also on his first long-duration space-flight. Turned out Barratt was having sudden issues as well.

"We called it down to the flight surgeons on the ground, and they were very concerned," Thirsk said, "so they wanted to investigate this. They quickly developed a protocol for us to use the on board ultrasound machine, which we normally used for life science research, and to image our eyeballs."

The crew had never performed this procedure before, but they had enough generic knowledge to figure it out — plus, people on the ground talked them through it in a process NASA likes to call "telementoring," something Thirsk was familiar with from the work he had done preparing for NEEMO-7. The astronauts ended up getting "excellent visualization" of the back of their eyeballs, Thirsk recalled, which showed the source of the trouble: the back of the eyeball was swelling, as well as the optic disk that leads out of the eyeball and into the brain.

"So everyone was really concerned now," Thirsk said. A shuttle was on its way to orbit in two weeks, carrying Payette and the rest of her crew, so NASA worked overtime to get a new piece of eye equipment onto the orbiter just before its spaceflight. Usually it takes a year or two to get the paperwork and safety checks done, but under these emergency circum-stances, countless number of people on the ground finished the necessary steps in days, so the electronic ophthalmoscope could fly to space. It's the sort of work that few people remember when talking about space missions, unless it's a famous emergency like the Apollo 13 explosion in 1970 — the thousands of team members on the ground who solve problems and make things happen so that the work in space can carry on. The person shepherding much of this work was flight surgeon Doug Hamilton, Thirsk said, who was "heroic" in leading the effort to lift the eye equipment off the ground.

In any case, the new equipment confirmed the problem: severe eye swelling that doctors call papilledema. NASA sent up new eyeglass prescriptions for the affected astronauts, but the prospect was so troubling that almost immediately after the astronauts came home, they were sent for MRIs. The pictures coming back were extreme — a tortuous and swollen optic nerve sheath, not to mention that the back of the eyeballs were flattened such that their focusing distance (or focal length) was altered.

It's been more than 10 years since Thirsk first squinted at that checklist, and it turns out that he and Barratt are far from the only ones affected. Thirsk knows of at least a dozen other astronauts who faced similar issues in orbit, and you can bet NASA is checking out every possibility. Perhaps the newly installed ARED on Thirsk's mission contributed, since no astronauts reported issues beforehand and ARED allows people to do more rigorous weight training, which could temporarily raise blood pressure. It could be due to elevated levels of carbon dioxide in the cabin atmosphere, or the weightless environment. Many studies have been published, and NASA is investigating as best as it can, but no firm answers are forthcoming yet.

We are no longer space visitors, and like any long-term resident, we are discovering the joys and troubles of living in a place for a long time. Things break, medical issues come up, there's the occasional bout of loneliness or depression that must be managed. But if there's a spot to figure out these issues in space, it's the ISS. Best to know this stuff now before venturing off on a two-year journey to Mars, or even deeper into the solar system.

CHAPTER 7

Nine years an astronaut

Courage and perseverance have a magical talisman,
before which difficulties disappear and obstacles
vanish into air.

— John Quincy Adams, Oration at
Plymouth (December 22, 1802)

With the smoke lingering on the Baikonur launch pad, I anxiously moved towards the stands. There, I was looking for anybody with the Canadian Space Agency who might have a radio or some way of knowing how Saint-Jacques, McClain and Kononenko were doing during their eight-minute journey to space. Eventually, I tracked down somebody holding a walkie-talkie-like device that was reading out the milestones in Russian. Beside us, our interpreter — Elena — mimicked what the astronauts and mission control were saying. Altitude milestones. Short observations about instruments. Then the magic words: "Engine cut-off."

It was Dec. 3, 2018. For the first time in five years, a Canadian Space Agency astronaut was in space. Safely. And on his way to the International Space Station. Barely six hours after his epic liftoff, about 150 of us crowded into an old Soviet Union theatre in Baikonur. We applauded when Saint-Jacques came out of the hatch of his spacecraft and greeted the three humans already waiting for Expedition 58 at the space station. Canada's Governor General — Julie Payette — said a few words in Russian, French and English. The astronauts briefly spoke with their family members, and then that was that — the broadcast finished, and we all went back to our hotels to pack for the long journey home.

In this book, I have deliberately ignored most of the activities astronauts do during flights, because these are well covered in the literature elsewhere. I've always believed that the time on the ground is not covered enough. To be sure, I understand the stark realities of journalism resources — I graduated not long before the 2008–09 recession and saw several workplaces affected by it. I edited this book in May 2020, as journalists around the world once again faced layoffs and workplace hours reductions due to the novel coronavirus pandemic. I also know the brutality of the news cycle that demands reporters constantly generate content on a variety of topics. But over time, this focus on the spaceflights leads to the public perception that all an astronaut does is get ready for flight and be in flight, and anything they might do during their long decades of work on the ground is incidental. Another nasty myth is what I like to call the "solo astronaut" phenomenon, where all of the attention goes on the astronaut and little (or none) to the family, friends and colleagues supporting their work.

So let's move back in time. Let's look at how an astronaut begins becoming an astronaut. As Hadfield cleverly and repeatedly pointed out in *An Astronaut's Guide to Life on Earth*, there is really no defined moment when one starts training for astronaut-hood. It happens long before you're selected, and some astronauts I've interviewed were barely finished being toddlers when they first wanted to go to space. (Remember

Sidey-Gibbons?) There is no one typical path in life for any professional astronaut — people joining the ranks include Navy SEALs, seasoned aviators and pilots and noted scientists. And for every Sidey-Gibbons, there is a Garneau, who just did interesting things without space in mind — until an astronaut application caught their attention.

You could spend a decade or more being very good at what you do and rising to the top of your field, but when astronaut selection time comes, you get cut — which is why every astronaut I've talked to (or at least, those thinking about space very early) emphasizes you have to enjoy the journey and the career, just in case you never make it to space. Joshua (Josh) Kutryk remembers getting to the last set of interviews in 2009 along with Saint-Jacques and Hansen, which he called "the bitter, bitter end."[1] This is after a year of countless exams, medical tests, simulated interviews, team-building sessions, emergency testing in fire and water and other hazardous environments, and who knows how much time off work travelling and preparing for these various opportunities. And then, because there were only two Canadian slots available in the 2009 astronaut class, Kutryk was cut.

Kutryk is one of those people practically born loving space, since he was interested "ever since I could remember . . . the biggest thing I remember is being curious and wanting to explore. That might have been on a bike, it might have been in the mountains, I just always had this sense of curiosity, wanting to explore."

You could fairly point out he was "only" 27 when he was cut, although a number of astronauts around that age have been hired — including Sidey-Gibbons a decade later. But still. To be in the last room and "then to be asked to go away" quickly turned an exciting experience into "a very disappointing experience," he said.

"It is a very good life-learning event, is what I would say," Kutryk added, and in hindsight, he said it did help him. He said the process gave him insight into what people are looking for when hiring astronauts, and while he was careful not to mention his "loss" to colleagues or to focus overmuch on being an astronaut, he did pick up a few certifications in the years following.

Specifically and incredibly, since that astronaut selection he earned three (three!) master's degrees in space studies, flight test engineering and defence studies from Embry-Riddle Aeronautical University, the United States Air Force Air University and the Royal Military College of Canada. And he continued his career in the Royal Canadian Air Force, ending up evaluating technologies and systems on Canada's famed CF-18 fighter jet to make it safer for pilots coming after him.

If that doesn't sound impressive enough, this person who grew up on a cattle farm was also a flight instructor and could fly more than 25 types of aircraft. In 2016, another astronaut recruitment began, and after another gruelling year of testing, Kutryk was in.

There's a flurry of media attention whenever astronauts are announced, and Kutryk and Sidey-Gibbons received an unusual amount in 2017 because they were unveiled on Parliament Hill during Canada's 150th anniversary of Confederation. (A space journalist friend of mine, Sean Costello, reminded me also of the following: Kutryk's younger brother, Captain Matthew Kutryk, was flying the "Canada 150" themed CF-18 Demonstration Hornet overhead as part of the opening for the mid-day event, just as his older brother was preparing to be introduced to the country down below.) The occasion itself was difficult for attendees, between pouring rain and strict new security requirements for visitors on the Hill, which included screening at two access points that held people in line for many hours. Still, it was a grand reveal — grand until Kutryk and Sidey-Gibbons had to quit their current positions and plunge into the reality of training.

NASA carefully calls its new recruits "astronaut candidates" because, like in the military, you need to pass basic training before you can fly. In the past (preparing for brief space shuttle flights), that training might have been a few months, but today it's a gruelling process lasting up to 2.5 years, depending on availability of instructors and facilities. The agency is usually spot-on with its selections, in that people selected for candidate training tend to pass. But that's not a guarantee.

Kutryk and Sidey-Gibbons were in a class of a couple of dozen people. Midway through learning about spacewalks and space systems, wilderness simulations and many more exams, one of their class, former SpaceX flight reliability senior manager Robb Kulin, left. His reasons for quitting were not disclosed due to privacy, but he was the first American astronaut candidate to quit training in 51 years, after John Llewellyn in 1967. (Other astronaut candidates have left the corps before Llewellyn, too.)[2] That's not a bad track record for such a difficult career. The Canadian Space Agency has also lost a few talented recruits before they made it to space — Money (who stuck it out for almost a decade after his 1983 selection), McKay (who resigned in 1995, after three years, due to a medical issue) and Stewart (who resigned a week after being selected in 1992 and was replaced by McKay).

Supervising the 2017 class was Canada's Hansen, who himself has been patiently waiting for spaceflight since his selection in 2009 and acting as a professional support to Saint-Jacques and his family through the long years of training. Hansen's role in supervising the class was a first for a Canadian and a reflection of how much NASA trusts his experience, chief astronaut Patrick Forrester told me during a brief interview in Baikonur in June 2018. Hansen has called himself a "den mother," working to make sure each astronaut candidate has support for the facilities that they need, and acting as a mentor to people who are just getting their heads around the fame, the travel, the long hours and all the professional demands that go along with being an astronaut.[3]

Astronaut Victor Glover remembers the ceremony at which Forrester announced Hansen's appointment to supervise the class. "He said, 'I put him in that position just because I think he embodies so many of the characteristics that are important,'" Glover recalled. "I remember thinking, 'Could you get a better compliment?' The head astronaut says that he puts you in charge . . . of training not just the two Canadian astronauts but the entire class, because of his respect for your character and personality and your intellect. I just, I remember thinking to myself, that is about as good a compliment as anyone could get. I've always respected him and just think the world of him."[4]

Conceivably, Hansen could be training people that fly into space before him, due to Canada's relatively small contribution to space station activities. Opportunities could, however, accelerate in the near future as commercial crew vehicles come online, and at least one CSA official has told me that they are pushing for Hansen to become a pilot — if possible — on one of these vehicles once Americans certify them.

But in the meantime, while Canadian media occasionally checked in on Saint-Jacques during his flight, Hansen's key role in mentoring a couple of dozen new astronauts is rarely mentioned. I asked him if it was hard to explain to media (and any other curious people who talk to him) what it means to be 10 years a potential space flyer and not assigned to a spaceflight.

"I do feel like it's harder on other people than it is on me, but because of the limited perspective," Hansen said.[5] "As an astronaut, you're part of a team that's doing incredible things. And it's not very hard to look at this and realize that even if I knew I couldn't fly in space, [why] would I not want to do this? This is still a really amazing opportunity. Just the things I do getting ready to go to space. And the team that I get to work with. Pushing technology for humanity's benefit. And these are things that I think are important anyway. And then the icing on the cake is I get to go to space."

So what are the things that Hansen does to get ready for space? Funnily enough, some of those roles are more famous than his job supervising the 2017 astronaut class. Periodically, Western University planetary geologist Gordon Osinski — affectionately known as "Oz" among the space community — will bring an astronaut along on a field expedition. This is not a ploy for publicity, as Oz is great at generating media on his own (and has rapidly grown Western's influence in space matters, including a separate degree that didn't even exist when Hansen and Saint-Jacques were selected). Rather, it's an opportunity for the astronauts to practise geology in a real-life environment. In fact, all astronauts and astronaut candidates periodically do wilderness training — in part to learn how to survive if their spacecraft goes down somewhere unexpected, but mostly to learn how to work in teams dangerous situations. One twisted ankle and your whole game plan changes. "These types of adventures, they will eventually put you in situations where you have to make tough decisions without knowing what the right [one] now necessarily is. We're looking for that skill set," Hansen explained.

The nature of what he does necessarily changes from expedition to expedition. With Oz's group, he takes time to mentor the students and make them aware of opportunities in space exploration. For higher-profile opportunities such as when he lived in the underwater habitat Aquarius for NASA's NEEMO project he passed along positive messages about the environment.

"I realized that there's weather down there," Hansen said. "Some days the current is strong. Some days the visibility is better. Some days it's wavy. You can feel the waves even when you're on the ocean floor, in your ears, when it's wavy. And it's just all of these things. The habits of the fish were just remarkable to watch, that the same fish do the same things every day. You could set your clock to it."

Every mission is an opportunity for data gathering, he said. While it's easy for journalists to report on astronauts in scuba suits doing sojourns outside of their habitat — would those be called "swimwalks," I wonder? — what the astronauts are doing is far from play. They evaluate systems that could be used on the International Space Station, such as time trackers or virtual-reality headsets that offer "just in time training" so astronauts can relearn old skills on the spot. For Hansen's mission, the team was also evaluating how a time delay would factor into mission planning. Some future astronaut team on an asteroid or on Mars will be grateful that NASA examined time delays now, although the payoff will likely not be for decades.

Canadian astronauts also take shifts in mission control, which forces them to have good systems knowledge of the International Space Station to communicate properly, Hansen said. It also shows astronauts close-up how the ground team makes decisions. It's a lot more conservative than what you would expect at first, he said, because safety is uppermost in everyone's mind. A newbie astronaut would not understand this as well if they were thrust into space without spending time among these controllers, he explained.

"If you put yourself in the place of an astronaut on board and you ask a simple question, sometimes you may not get an answer as quickly as you would think — since it's a team of experts sitting on the ground who, in theory, this is the only thing they have to do that day. And so you could be left scratching your head wondering why it's taking so long to get this."

Ground astronauts also act as supports for their teammates in orbit. This process can actually start very early in their astronaut career. Hansen had only just completed his astronaut candidate training when Hadfield went into orbit in December 2012. He acted as Hadfield's casualty assistance and calls officer (CACO). On the one hand, this means if an emergency happens in space, Hansen steps in to take care of the family. (MacLean took on this role during the fatal 2003 Columbia accident for Ilan Ramon's family.[6]) But it also means that Hansen was perpetually on-call for Hadfield during the astronaut's time in orbit. This meant assisting not only Hadfield but his family with anything they might require.

"The expectation was he could just reach out to me and ask me to do things that would help relieve his schedule, and help him get to space, without a load on his mind," Hansen said. "And so I just did all those things. And then sometimes changing the brakes on his car, or talking to [his wife], Helene." He made reference to the fact that I interviewed Helene during Hadfield's flight in 2012–13 and added, "You've met Helene; she's very independent and self-sufficient. It wasn't a hard job, for sure, doing that for Chris. There weren't many things asked for, but just some things like that. Which was great because it just gave me more access to Chris. More mentoring from Chris as well." Thus, there were strong benefits to this extra responsibility for Hansen. Hadfield was not the only astronaut he was helping at the time. He served similar duties for Kevin Ford, who launched two months before Hadfield, and Hadfield's crewmate Tom Marshburn. Generally, the CACO role goes to a qualified astronaut, so when Saint-Jacques flew in space, astronaut candidates Kutryk and Sidey-Gibbons were not yet ready to take on the role. (For those who are curious, the CACO role fell to Payette,[7] a long-time friend of the Saint-Jacques family — another reason she was on site in Baikonur, in addition to the diplomatic opportunities for Canada.)

But still the younger astronauts participated. Back in the Montreal area, Sidey-Gibbons led delegates through a televised broadcast of the Saint-Jacques launch, while Kutryk acted as a helper for the Canadian delegation in Baikonur. Most of Kutryk's role was akin to that of a best man at a wedding, in that he was helping with all the little guest-logistics matters that come up. This human-relations role is probably as important to an astronaut candidate as the official reason he flew out there — to be exposed to the vast launch operations at Baikonur.

"It is very different. You know, their language is different, there are cultural aspects that are different," said Kutryk, who, two years into his Russian training, was likely somewhat proficient in the basics but realized there was a long way to go.[8] "The technology is different," he added. "The geography is very different. And so it puts a lot of context into the subject matter. You can only read about that stuff so much, and it really helps to go over and to see it, and to just see how good of a job they do at launching humans into space."

The orbiting astronaut racked up several milestones during his 204 days in space, particularly when the first Crew Dragon spacecraft arrived at the space station on March 3, 2019. SpaceX's Crew Dragon — along with Boeing's CST-100 Starliner — are the two next-generation vehicles astronauts will fly very shortly.

It's taken about a decade of development to get these spacecraft ready after the space shuttle retired in 2011, because space is hard and funding for contracts was at times limited. So it was incredible to see this future astronaut transport arriving — and even more incredible to see a Canadian-wielded Canadarm2 gently snatch it while it was station-keeping, or waiting nearby the International Space Station.

There's a whole international team that takes care of the robotics on station; in Canada, there's even a separate robotics mission control at CSA headquarters in Longueuil, QC. Overseeing the robotics (and in fact, the whole Canadian mission plan) was Podwalski, the Canadian program manager for the ISS program. Long before Saint-Jacques carefully picked up Crew Dragon, Podwalski was puzzling over how to give Saint-Jacques enough time to do this — and all the other tasks that space station crews are assigned.

You see, when Saint-Jacques launched on Expedition 58, the space station was at an unusual lull in crew size. The abort in October — the one that had me lying awake in worry on Saint-Jacques's launch day — precipitated a chain of temporary staffing issues in orbit. The aborted Expedition 57 was supposed to carry two people into orbit, but clearly they never made it there. This disrupted the usual cycle of space missions. After a new crew flies into space, they work together with the existing crew on the space station for several months. Eventually, the old crew flies down, the new crew works alone for a few weeks, and then more people come up to continue the cycle.

Not this time. Expedition 57's abort on Oct. 11, 2018, was so serious that for weeks it wasn't clear if or when the Expedition 56 crew would be relieved. The space crew immediately told NASA they would stay up as long as needed, but the international partners considered several scenarios (with the most radical being bringing the current crew home and leaving the space station empty for the first time in two decades, although it was only brought up as a vague possibility).

So things were uncertain. For weeks, Saint-Jacques's December launch date seemed likely to be delayed, but after the Russians quickly tracked down the issue, Saint-Jacques actually launched nearly three weeks *early* to the surprise of many. The reasoning was bringing Expedition 58 up a little sooner would give precious time for the two crews to work together before the current crew's scheduled return.

But here was the rub — once the three astronauts waiting for relief in space departed, Saint-Jacques and his Expedition 58 crew were by themselves on the space station for three months. (Generally, space station staffing is only at a minimum for perhaps three weeks.) And unfortunately, in these situations, science is usually the first thing to be cut. You can't delay space station maintenance, and you don't want to mess around with the timing of supply ships either.

Just ask NASA's Scott Kelly, who in his year in space saw so many supply ships coincidentally destroyed or misrouted while flying to the space station that his crew was threatened with needing to ration food and other supplies. (That never came to pass, fortunately, and the cause of each failure was diagnosed and addressed for future flights.)

So when it came to Expedition 58, Podwalski and his international team had two main challenges: how to design a new flight plan in six weeks, and how to preserve priority items for Saint-Jacques and his American crewmate, Anne McClain, in taking care of the American and European segments of the space station. And many discussions went on with Roscosmos, who is responsible for the Russian side and only had one astronaut there for many months. Fortunately, NASA told Canadians early on that all of our experiments would be preserved — a nice gesture, given that Canadians only get to fly every few years right now.

"They know that international partners don't get the same opportunity or the same volume of opportunity that they do," Podwalski said.[9] "So they really do want to make sure that . . . those partners come away happy with what they get in terms of the return on investment, if you will, in terms of the science return."

It turned out that McClain and Saint-Jacques performed their work with gusto, completing tasks at a higher rate than people anticipated and slightly making up for the shortfall in staffing. But changes had to be made still. Other (non-Canadian) experiments were shuffled, and it was decided to wait on the three NASA–Canada spacewalks planned for the Expedition 58 crew until more astronauts were on board, Podwalski said. That's not only for practical reasons but also for safety reasons. In a three-person crew, a spacewalk would mean that only one person is on board the space station while the other two work "outside." It would be unsafe to send one person out alone and leave two behind, for reasons that any Canadian who has been in the wilderness would appreciate. At the time of our interview (February 2019), the new crew still hadn't flown up to join Saint-Jacques and his companions, so the timing of the spacewalk was uncertain. But the new crew arrived safely, and the spacewalk went ahead the following month. Later, Saint-Jacques became the first CSA astronaut to float outside since Dave Williams in 2007. Despite the 12 year interval, Saint-Jacques called Williams two days before the spacewalk to discuss one of the tasks. Williams, who had been the last astronaut to work on that part of the space station, was stuck in downtown Toronto traffic while Saint-Jacques overhead was travelling 25 times the speed of sound. Such is the nature of spaceflight.

And Saint-Jacques also manipulated Canadarm2, adding to a long line of proud Canadians who have been able to use their own technology in space. Some of us can name the astronauts who have done this. But very few can name any Canadians who manipulate robotics from the ground — in fact, it's probably not well-known that we have a robotics mission control right here in Canada. Podwalski said at the beginning of the program no one imagined a small international partner would receive that much independence and trust from NASA. But over the years, Canada continued to show its worth — and NASA came to appreciate how delegation could save astronauts valuable time. After all, if ground controllers are able to do a share of the robotic work, that's less work for busy astronauts to worry about.

Podwalski discussed how the relationship with NASA changed. "[Years ago] they'd never believe that we were going to be able to get to that level, and honestly, there was a bit of the turf war [that] kind of mirrored the operations community that runs things in space." He added that NASA was not "necessarily convinced" that Canada could work on its own, but as a partner, Canada did what it is good at doing — "it stepped up in every possible way, in terms of building all the credibility."

Over the years, the role of Canada's ground robotics team has evolved into something like an apartment landlord. If something breaks on the outside of the space station, the partners eventually need to make a decision — to send an astronaut outside or to address it using robotics controlled by the ground. So, for Podwalski, this means a fair number of late-night phone calls when the latter is being considered.

"We get called in the middle of the night when there's a bump, and somebody heard something on the station, and they suspect that there may have been some kind of micrometeorite strike," he explained. "Because we're effectively a movable camera on the outside of the vehicle, we're usually in that first line of, you know, 'What if we want to go and check something out?' We can do that with the robotics."

Robotics is also perfect for one of those scenarios that keep mission managers very worried: ammonia leaks. This toxic coolant is essential for keeping space station modules and their experiments from overheating, and from time to time, leaks outside do occur. When astronauts go outside to address the issue, many safety precautions must be considered.

It's common practice, for example, to make astronauts wait for an extra while outside to make sure the sun "bakes" off any ammonia flakes that could have landed on the spacesuit. Robotics operations don't have to contend with the same issue, Podwalski said. Dextre is equipped with a "sniffing tool," or an ammonia-leak detector, that can characterize the seriousness of a leak without astronauts needing to think about pulling on a spacesuit. This buys the international partners a bit of time in considering their next move.

And emergency operations are only a fraction of what the robotics operators do, of course. Dextre is now exclusively controlled by people on the ground, and operators routinely do things such as pull equipment out of the cargo bays of visiting vehicles, move items around on the outside of the space station, and take routine images to make sure the space station exterior isn't getting too damaged from micrometeorites and radiation — the two chief threats that space throws at equipment.

Shireman, who manages the ISS program on the NASA side, has high praise for the ability of Canadian roboticists to do all this work from the ground — without a single moment of astronaut time required — while the Canadarm2 (which astronauts do operate) was an essential part of space station assembly. In fact, Canada was put on the "critical path" for assembling the space station, which was a symbol of trust in our country's capabilities.

"The expertise that Canada has brought to the table both in the shuttle program, down to the International Space Station program and ultimately now to the Gateway program, has been spectacular," he said.[10] "We could not assemble an International Space Station without robotic technology that the Canadian Space Agency has brought forward. There's just no way."

This technology has been especially helpful as the ISS transitioned into an operational phase, a phase in which NASA is pushing as hard as possible (while keeping safety requirements in mind) for astronauts to use their time for science instead of maintenance, especially on the exterior of the space station, an inherently dangerous place to go. Dextre's robotic operators can shorten spacewalk time by moving wire coils, robotic foot adapters, ammonia tanks or other pieces into place for astronauts to pick up during spacewalks — and do checks for leaks or damage on the exterior without needing to bother the astronauts.

"It's really been beneficial not only to Canada and the United States, but really the entire partnership," Shireman said. "Our astronauts work on experiments. They don't work on flying the robotic arm. We have people here on the ground that fly the robotic arm."

Podwalski added that — no surprise — such coordination

on robotics and other mission planning requires dozens of meetings with NASA for robotics teams, logistics teams, hardware teams, system engineering teams, the folks that handle telemetry (or data gathered from instruments in space) and even the contractors.

Podwalski estimates that more than two-thirds of the 75 CSA employees directly involved in space station planning "are probably tied into various boards and panels with NASA and our partners." This planning is just one example of how the partners can look past political and international differences to keep the complicated space station going, he added.

For Saint-Jacques's mission, NASA handled the day-to-day planning and routine maintenance, but Canada did have influence on which Canadian experiments ran, and how and when he communicated with reporters and students on the ground. The planning of the mission, of course, took many years and involved such tasks as selecting which experiments to run, ensuring the needed equipment was on board the International Space Station and interfacing with the principal investigators to decide how often and when they would collect data.

It has already been pointed out that Canada has 2.3 percent utilization of the space station, but that's not a constant. It tends to swell when Canadian astronauts are on board and then to fall back when those astronauts return, Podwalski said, although it evens out to 2.3 percent over many years.

This is just a fraction of what goes on during Canadian missions when people aren't watching, but part of the CSA's job is to try to get people watching as much as possible. And that's a lot easier today than it was in the days of Garneau, when people tended to get their news from newspapers and nightly television broadcasts. Like every adept modern organization, the CSA leverages social media — Facebook and Twitter — to get the message out. And everyone of a certain age knows that the social media story really began with Hadfield.

Hadfield became world-famous practically overnight on Jan. 3, 2013, when Canadian *Star Trek* actor William Shatner sent him an incredulous message: "Are you tweeting from space?" Hadfield's careful answer was so on-message for Trek fans — "Yes, Standard Orbit, Captain. And we're detecting signs of life on the surface" — that the exchange quickly went viral and a number of other *Star Trek* actors joined the conversation.[11]

That could have been a flash in the pan, but Hadfield, along with his family and the CSA, worked together to create a social media campaign to highlight the role of Canadians in space. Wide use of social media was brand new in 2013, so even little things got a lot of attention — Hadfield showing how to cut his nails in space on YouTube, or wringing out a wet wash-cloth of water and watching it ooze for the entertainment of schoolchildren (the result of a national contest to pick a science experiment to run in space). And a lot of background negoti-ations went on for some of Hadfield's bigger ventures, such as playing David Bowie's "Space Oddity" in orbit — this required permission from Bowie and his label, as well as a team on the ground to splice the music video and sound together.

I remember a few folks complaining that Hadfield was distracted from his mission by doing so much social outreach, but those people didn't get what was really going on. First of all, Hadfield was doing all of this work in his limited spare time — and more importantly, Hadfield was supported by CSA's media-relations team and several family members (most notably, his son Evan) in getting the word out. Also: Hadfield actually was scary-productive behind the scenes. On the day of his May 2013 landing, Saint-Jacques proudly told me that Hadfield had run the most scientifically productive mission on the space station to date. And that included dealing with an emergency ammonia leak just days before landing.[12]

As I write about the behind-the-scenes folks supporting Saint-Jacques's flight, I'd feel remiss in this book about Canadians in space if I didn't mention the Canadians in space that no one knows about. Undercover astronauts, you may be asking? Nope. I mean *undercovered* astronauts.

Take this example: when Saint-Jacques took his "walk" outside of the spacecraft, several Canadian media (including me, in a hastily scribed tweet) erroneously referred to him as the first Canadian to walk outside the spacecraft in 12 years.

In fact, I heard that mistake repeated in a French-Canadian radio broadcast months after the spacewalk, so it sounds like this error is in danger of being written into history textbooks. Let me be very clear here: he wasn't the first to walk outside in 12 years. In fact, a Canadian had walked outside barely a year before, in 2018 — Andrew (Drew) Feustel. But hardly anyone knew about him in our country because his home agency is at NASA.

While you're still picking your jaw off the ground, understand there have in fact been *two* Canadians in space outside of the confines of the CSA. NASA accepts Americans and dual citizens, and given its agency flies far more often, the temptation is there for Canadians to try that route. To be sure, this isn't akin to the brain drain of the 1990s or the "war on science" that drove scientists south and abroad when money and government policies seemed attractive elsewhere. Two Canadians out of the 561 astronauts who have flown (as of this writing) comes out to a fraction of 1 percent. And it's not as though they would have had a better chance in Canada, as our country easily fills its astronaut slots at the time of each selection.

I asked the two Canadians, Gregory Chamitoff and Drew Feustel, to comment for this book. Chamitoff initially said yes, but didn't respond to multiple requests to secure an interview time, so I pulled what I could from media reports. He certainly wasn't talked about much in Canada during his 2008 flight, except in his home city of Montreal. One of the top Google results I obtained was about how he wanted to bring that city's world-famous bagels into space with him,[13] which is a great human-interest story, but shows the low level of public discourse going on about his flight in Canada at the time.

While Chamitoff was born in Montreal, the space-obsessed kid (who remembers loving it at age six) migrated to the United States as a young child. "My dad always felt that California and the United States were the land of opportunity, so as soon as there was a possibility to come, we came. My father already had family and my mother had family also in California," he told NASA during a pre-flight interview in 2008.[14] Despite accumulating an engineering-heavy resume in multiple disciplines with stints at the Draper Laboratory, the Massachusetts Institute of Technology and the California Institute of Technology, it took Chamitoff several tries to make it into the astronaut corps, he added. But from there, his career literally rocketed off.

He was the first Canadian to do a long-duration mission aboard the International Space Station — yes, before Thirsk (2009) and most certainly before Hadfield (2012–13). Chamitoff was there for six months during Expeditions 17 and 18 in 2008. Then in 2011, he flew on the second-last shuttle flight ever — STS-134. This Canadian has racked up a lot of other milestones that people would appreciate, including operating the Canadarm, operating the Canadarm2 and doing two spacewalks — including the last spacewalk of the shuttle program. And his influence still lingers at mission control — he helped develop the screen display of the ISS and the space shuttle, starting in the early 1990s.[15]

Feustel became a Canadian through marriage. He met his wife at Purdue University when she came to finish her undergraduate studies and pick up a master's. The two of them moved to Belleville, ON, for a few years so that she could be near her family, and he ended up (on the suggestion of his Purdue advisor) attending Queen's. (Among other milestones, Feustel deployed seismic equipment at SNOLAB — the Sudbury neutrino laboratory that looks for fundamental particles to better understand how our universe was made.) Eventually, they migrated to Houston when he got a job at ExxonMobil, but by then he had Canadian citizenship and two children born in Canada who were natural citizens.[16]

His involvement in the astronaut program came through a series of chances. The first was living in Houston, which is where NASA's astronaut central — the Johnson Space Center — is located. Another was an acquaintance with Stewart, the Canadian who was selected to join the CSA program in 1992 but turned it down a week later. Stewart appeared on Canadian show *W5* around the time of his selection, and Feustel (who saw him on the show) recognized him five years later at a geophysics conference in Houston.

Stewart introduced him to fellow 1992 astronaut classmate Hadfield, who was fresh off his trip to Mir and happy to give Feustel some advice. "We became fast friends," Feustel said of his conversation with Hadfield, and in 1999, Feustel applied successfully for the astronaut class announced in 2000. He credited Hadfield with helping him succeed, as through their connection, Feustel met many people in the program, and by extension, many people got to know him before selection. Going to the CSA was not an option at the time, he pointed out — the last selection was in 1992, and no one knew when the next one would be. (It ended up being 2009, an incredible 17-year span.)

Feustel's space resume is every bit as impressive as those of the Canadians you've heard of. He worked on the Hubble Space Telescope — in fact, he was among the last people to ever touch the orbiting observatory during a big mission in 2009, around the same time the CSA announced Hansen and

Saint-Jacques as the next astronauts. Few knew that a Canadian was then working in orbit. In 2011, Feustel flew again in space — on the same flight as fellow Canadian Chamitoff, a fact that Canadian media largely ignored (even though they fussed over Payette and Thirsk meeting in space in 2009). And in 2018, Feustel made his own six-month journey aboard the ISS.

Feustel commanded the space station — just like Hadfield — and got to do three spacewalks. Along with his Hubble work, Feustel is one of the top all-time spacewalkers in terms of time accumulated "outside" — 61 hours and 48 minutes across nine spacewalks.[17] Incidentally, it seems the best background for a spacewalker and space station commander is not only being a high-level scientist or engineer but also a garage mechanic.

Feustel said that ability was essential for him: "All of that hands-on experience that I have on Earth working with tools, working in the garage, working with my hands . . . my career field as a field geophysicist, that really honed my skills of working in difficult situations, difficult physical environments, working with equipment. And that's similar to what we do when we're in space."[18]

Another thing that every mechanic knows is that something always goes wrong during a job. Just ask the guys who installed my new water heater the morning I wrote this draft chapter in mid-2019 — due to one missing anode, they had to drive back to the warehouse and add an hour-long round trip to their work. During a spacewalk, adding an extra hour is sometimes possible, but not always. There's only so much oxygen in the tanks. There's only so much energy astronauts have.

During the Hubble repair, fellow astronaut Mike Massimino found himself puzzled when a handle (on top of an access panel) refused to come off the old telescope to provide access to electronics inside. Massimino waited outside for about an hour while the entire crew (including Feustel) troubleshot the problem, with NASA concluding that because no astronauts were due to come back, it would be safe to simply rip off the handle — as long as Massimino was careful not to tear his suit on the jagged edges.

That's the most famous example, but Feustel said the crew encountered many such situations while they were working outside under time pressure: "There is always something that will test your ability to adapt and modify your expectations of scenarios in space to what the real scenarios are," Feustel said. As a rookie astronaut then, he admitted, "that was something I didn't fully expect."

Feustel has high praise for Canada's investment in robotics, because he's one of the few of us who have had the unique experience of riding on a version of the Canadarm *and* operating it. During the Hubble mission, crewmate Megan McArthur carefully directed him ("flew" in astronaut parlance, because what else would it feel like?) towards the telescope on the Canadarm's end effector. Then he rode Canadarm2 during the STS-134 space shuttle mission for a space station spacewalk.

As Canadians, we are trained to think this is extraordinary — here's a Canadian riding robotic arms, repairing space telescopes and taking part in the second-last shuttle mission, oh my — but Feustel has a different take. He pointed out it's more than likely that one will use the hardware at some point during a mission because it's so essential. Canadarm2 still snags cargo ships and plays a key role in spacewalks, and the planned Canadarm3 will do remote repairs on a lunar space station.

"This is hardware that supports our missions," Feustel said, which is both praise and a warning for our country — Canada is an essential part of the space station and needs to keep developing robotics to stay with the next generation of astronauts. And while I've been critical of Canadian media (which includes myself) in undercovering Feustel's missions, he graciously says the CSA has been as inclusive as possible of him given that he's a Canadian NASA astronaut. He did a "downlink" event — a call from space to ground — with the US embassy in Ottawa, bedecked in a Canada/US flight suit. Better yet, while in space, Feustel "called" Canadians during a national celebration; he was dialled into the loudspeaker on Parliament Hill on Canada Day 2018, just one year after Sidey-Gibbons and Kutryk made their debut as astronauts.

"I think it's been difficult for the Canadian space program to get behind me being in space as a Canadian, because I think it can be a distraction to the limited resources that they have and the limited astronauts that they have, that they call Canadian Space Agency astronauts," Feustel said.

"But we were happy that there was some support recognition. After all, I think that the key to anybody being in space for any nation is to inspire the youth and the young generations of that nation to think about science and technology and their impact on the world and what their future can be. So it was nice to see Canada get behind some of the activities that we did in recognizing that I can be not only an American, not only a US astronaut, flying as a US astronaut, but representing Canada and trying to inspire the youth across the nation. So we had a couple of good outreach events that were really quite enjoyable for me, and I, hopefully, made an impact on the kids in the nation in support of the space program."

That's usually what every astronaut comes back to in lengthy conversations — the fact that they see their true job as outreach, especially to youth who are at that influential age when they are figuring out what they are interested in. So the two NASA Canadians have done their bit. In that vein, why so little coverage of Chamitoff and Feustel by Canadians? I believe it's a problem of resources.

Few reporters cover space, and even fewer understand the distinction between CSA astronauts and NASA astronauts who happen to be Canadian. Astronauts are also a modest and polite bunch, so they don't tend to talk out much about media coverage even when they're retired — presumably because it would make them sound like they are more eager for fame.

For that reason, I really do admire Feustel's wife Indira, who will call reporters out on social media if they don't mention a Canadian milestone correctly. She keeps us accountable, even though that shouldn't have to be her job. (To be fair, she also celebrates good coverage of her husband and is an enthusiastic partner in preparing for events. My favourite example of this was when she asked Twitter if she and her husband should wash out the smell of space from one of his flight suits; the consensus seemed to be no.[19])

And to be fair, every reporter's resources get stretched, and we have to make decisions about what to cover and when. Freelancers, like myself, also must be wary of income-draining projects like those involving travel or covering events for your own blog (when no other outlet will pay for it), because in the long run, your income-producing projects must keep you afloat. So sometimes, unfortunately, the money comes first. Not only did I undercover Feustel and Chamitoff, but I had to turn down the chance to cover Saint-Jacques's return to the CSA after his flight. I had too many unrelated stories — and this book, and a house repair — to work on that day.

I have to make tough choices a few times a year — and I always regret that I can't be in two places at once. But, I remember, I haven't even hit age 40 yet, which means I have a long way to go. It's my hope that as my career progresses and I finish "adulting" responsibilities like paying down the mortgage, I will have the freedom to tell more of these Canadian astronaut stories to people. Hopefully this book — to paraphrase Armstrong's words on the moon 50 years ago — is a small step in the right direction.

CHAPTER 8

"Don't let go, Canada"

> The prince who relies entirely on fortune is lost when it
> changes. I believe also that he will be successful who
> directs his actions according to the spirit of the times,
> and that he whose actions do not accord with the
> times will not be successful.
>
> — Niccolo Machiavelli, *The Prince*
> (translated by W.K. Marriott)

Ottawa is a government town, and I use the word "town" quite
deliberately because the locals like to say our city is too small
and everyone knows each other — sometimes through bizarre
connections. My dad once travelled across the country and
met somebody who knew a friend on the dead-end street he
lived on. I, a space-crazy young adult, used to work at a tiny
community newspaper that happened to employ a Canadian
astronaut's brother. I also once interviewed a fellow who ended
up being my brother's boss about two years later. We all have
stories like that here.

In the latter weeks of February 2019, Ottawa felt even
smaller than usual. Everyone was literally watching the same

thing — even more incredibly, for hours on end. This was no repeat of the Kennedy assassination or the moon landing or 9/11, however. It was fresh testimony from a young Indigenous member of Parliament, Jody Wilson-Raybould. And when I say everyone was watching her, I do mean everyone. I would walk into interviews with high-ranking government officials, and every TV in the office would be on, watching her.

Wilson-Raybould has an incredible resume that includes being the first Indigenous person to be minister of justice and the Attorney General of Canada, positions she held for four years shortly after she assumed federal office in 2015. She had just taken on a new position as minister of veterans affairs in January 2019 when weeks later, a report came out in *The Globe and Mail*, a newspaper that everyone in Ottawa reads.

The Feb. 7 report out of the national newspaper's Ottawa bureau said that according to anonymous sources close to government, the Prime Minister's Office allegedly tried to influence her office's prosecution of SNC-Lavalin, a giant Montreal engineering company facing criminal charges related to several of its contracts.[1] (Coincidentally, SNC-Lavalin was also behind in delivering a notoriously unreliable light-rail transit line to the City of Ottawa, so you can imagine the local media jokes.) Trudeau and others close to him repeatedly denied the accusations, even as Wilson-Raybould voluntarily resigned her post and gave testimony at the House of Commons justice committee late in the month.

At 6:10 p.m. on Feb. 27, after a full day of testimony from Wilson-Raybould and non-stop coverage of her words across Canada, a message from the Canadian Space Agency flitted into my inbox. Sent on behalf of the Prime Minister's Office to a media list, the email had an agenda of events for the following morning, beginning with a 9:30 a.m. stop at CSA headquarters in Longueuil. "The prime minister will tour the Canadian Space Agency, and make an important announcement," the email said.

There wasn't much time to speculate as to what was going on, but I remembered NASA administrator Bridenstine visiting Ottawa a few months before and asking very directly, before a conference room filled with perhaps a thousand people, for Canada to commit to NASA's Gateway space station project at the moon. Besides which the Trudeau government had been promising for years to release a long-term space plan to help Canadian industry know on which sectors to focus. But unfortunately, I couldn't be there in person. I had somebody important to meet.

It was bitterly cold that Thursday morning in Ottawa as I parked at city hall and strolled a couple of blocks to a breakfast diner, recording equipment in hand. Minutes after my arrival, a man in his late 70s walked through the door, dressed in the trademark suit and heavy black coat that federal Canadian politicians favour. He was Kelvin Ogilvie, a recently retired Nova Scotia senator with a unique skill set in biotechnology, genetic engineering and bio-organic chemistry.

Born in rural Nova Scotia in 1942 during war years, Ogilvie attended a two-room schoolhouse in Summerville, on the banks of the Avon River. His grandfather was a sailing vessel captain, and his father enjoyed fishing in the river. His mother had obtained a teaching certificate from a teaching college. Ogilvie remembers inheriting a sense of exploration from his dad and a love of learning from his mother. The oldest of four kids, he was studious — the benefit of such a small school was you could learn at your own rate. "When I'd be in, say, Grade 2, I was in Grade 3–plus math books, you know, just to keep busy and not raise hell all the time," he said.[2]

In the middle of Grade 5, Ogilvie's family moved to the larger town of Bridgewater on the South Shore of Nova Scotia — an interesting location for the family given that Ogilvie could trace his maternal ancestry to the German-Dutch settlers of nearby Lunenberg in 1750. While the family was "economically challenged," as Ogilvie put it, both parents enjoyed reading and writing, and his father was also a published poet. By junior high, Ogilvie was reading *Life* magazine's dispatches about the war effort and NASA's early rocket experiments led by Wernher von Braun, a captured German with a Nazi background.

Repeatedly, Ogilvie felt the pull of the larger world, but he felt he couldn't fit in as a sailor. The alternatives weren't attractive to him: "I wasn't going to work at the service station for the rest of my life. There was nothing there for me that I could see, so university was just a natural instinctive next stage," he explained. Fortunately for Ogilvie, he enjoyed science, and that's where most of the scholarships were. He chose Mount Allison University in Grade 11 because "all the Grade 12 students that year seemed to be going to Mount Allison." Despite being a year younger, he got his scholarship.

Then God, in a sense, intervened. While Ogilvie's family was never deeply religious, they did attend the Baptist Church and were well-known to some of the senior people in the community. "I guess some people seemed to think I had potential," Ogilvie said, because before long those folks had leveraged a competing and more attractive offer in the 16-year-old's eyes: a matched scholarship at Acadia University, which had a deeper background in science than Mount Allison. Moreover, it turns out that his mother's family had lived there until the British expulsion of the Acadians forced them to the Summerville region. So Ogilvie felt a family tug towards that region.

"We took a drive down, and it was very picturesque. So I made the decision that I would go to Acadia," Ogilvie recalled. Ogilvie fit in perfectly, swiftly choosing chemistry as his major and graduating with honours. He felt too impatient to take the master's–Ph.D. route offered at the leading Canadian universities of the day and turned down the chance to be nominated for a Rhodes Scholarship out of a fear that the British universities were not innovative enough.

There was no trouble looking for financial aid in the United States, however, because his honours chemistry completion gave him automatic acceptance to all the Ivy League universities (a funding situation that would be envied by today's students). Moreover, Ogilvie recalled, more than half the heads of biology departments in major US universities had an Acadia undergraduate degree because its life sciences program was so renowned.

This was 1964. The United States government was living in fear of the Russians, who seemed to be pulling ahead in the space race to bring humans to the moon by the end of the decade. Money was being poured into the educational system to create technicians, and other sciences were burgeoning as a result — including DNA research. Three male scientists had just received a Nobel Prize for their work solving the structure of DNA two years before; female scientist Rosalind Franklin's contribution was not acknowledged until decades later, although the Europeans recently named a Mars rover after her.

Ogilvie got his footing in science, under the supervision of Robert Letsinger who worked on chemical synthesis of DNA. Ogilvie created a new method to create DNA on a large scale. At the time, it was painstaking to make even a milligram or two of DNA. The first time he made 39 grams of dinucleotide, Ogilvie recalled, he was in awe.

"Dinucleotide was selling for $100 a milligram in a chemical catalogue. That didn't mean that that was worth $4 million, but it was a nice symbol, and we all joked as I walked around the lab with this bottle," he said. This was just the beginning of a storied scientific and policy career for Ogilvie, which included inventing the drug ganciclovir to fight infections in immune-compromised people, serving as president and vice-chancellor of Acadia for 10 years, and receiving the Order of Canada.

In reading this mini-biography, you can appreciate that Ogilvie is more steeped in science than the average senator. He has competed for research dollars, completed a high degree, made discoveries that affect everyday society, participated in the culture of science — a stark contrast from the political science, law and history graduates that fill the ranks of many political circles in Ottawa. While his expertise is not space, space influenced his career. And as we went through his star-studded resume, I thought again of the mysterious Trudeau announcement taking place literally as we spoke.

While that was on my mind, Ogilvie had moved on to his worries about science policy in Canada. For at least the last 25 years, he told me, "we have continued to fail relative to other nations of the world" — in large part because Canadian universities, while strong, "lack the character to realize that more is required of them than absorbing money from society." Average Canadians also don't understand why intellectual property is so important.

Ogilvie said he tried to do his part to influence policy. While never a career politician, he eagerly accepted an offer from the Prime Minister's Office in 2009 — just a year after a near-sale of part of MDA's space technology hit the media and caused much head-shaking in the space industry — because the Harper government felt it would be wise to add more science expertise, he was told. Perhaps the choice was curious — Harper was never known as a science-friendly prime minister — but Ogilvie leveraged the opportunity. For him, the highlights were modernizing Canada's legislation on prescription pharmaceuticals and helping to restart Canada's national science advisor position.

But despite this progress, he remains frustrated that there is no "coherent" national science policy. He pointed to space exploration. Yes, the rumour was that Canada was ready to fund a focused space program once again. But that's not an overall science policy, he said. Even the National Research Council, as hard as it works to match up emerging science with emerging industry needs, only covers a fraction of science needs in Canada, he argued.

While he praised the Canadian government for working on space policy as we met in that diner, he found the timing of the announcement, given Wilson-Raybould's testimony, distressing. "The announcement itself will not get the attention it should have," Ogilvie argued. "I'm sure the questions will be all about what's going on in the Hill yesterday, not on the benefits of the space program. And that's a tragedy. So there is another example of Canada having no concept of the value at the leadership level of a real science policy. This should be timed in a place and circumstance where there's an attention on science."

He added this wasn't the first time Trudeau's government had, in his eyes, stumbled on science policy. He cited reports he helped write recommending a national strategy on obesity, a national strategy on dementia, even a national strategy on artificial intelligence — the latter considered a key sector for Canadian growth, including in space exploration. Every time, he argued, the ministers were uninterested. The minister responsible for artificial intelligence, he added, wrote the most unenthusiastic response he had ever seen in nine years of serving as senator.

Ogilvie, to be fair, is just one voice in the science community in Canada — a powerful and informed one, to be sure, but not the only perspective on Trudeau's announcement. And it was indeed an influential announcement, promising that Canada would participate in the Gateway by contributing a "Canadarm3," a robotic arm equipped with artificial intelligence.

I asked Garneau, now the country's minister of trans-port, why the announcement came in the middle of all this controversial testimony. His response had two parts to it. First, he made it clear that it was a coincidence, because the schedules of all four active Canadian astronauts (including one in space at the time) had to be coordinated — not to mention, several retired ones made a guest appearance in Longueuil with Trudeau. The second part was more prosaic: "We're doing the business of government, and this is part of the business of government, and we will continue to do the business of government. We have a responsibility to Canadians."[3]

Ogilvie's prediction about distraction came true in some media. CityNews Toronto pointed out that the first question at the Canadarm3 press conference was not about space or robotic arms or astronauts but about Wilson-Raybould.[4] The *Montreal Gazette* ran a story that day about Wilson-Raybould's testimony, based on questions reporters asked about it at the epic CSA announcement. It described "a press conference that was supposed to be about interstellar exploration at the Canadian Space Agency in St-Hubert but was overshadowed by the SNC-Lavalin scandal."[5]

Not all media were that judgmental, though — especially those with more experienced science reporters. *The Globe and Mail* ran an online story fully focused on Canadarm3, only mentioning Wilson-Raybould in a "related" headline list midway through the text.[6] And CBC ignored it completely, instead focusing on matters such as how artificial intelligence would contribute to Canadarm3.[7] (That said, I did a series of CBC interviews on radio in the days afterwards, and CBC Ottawa's host asked me directly if I had asked about Wilson-Raybould while writing my own story — which I had — so at least one reporter or producer there was thinking through the relation.)

I asked Manley — a former Liberal deputy prime minister who, as you saw earlier in this book, took a large interest in CSA affairs as minister of industry — about the Canadarm3 announcement. His take was actually quite apolitical and focused more on the financial side: "I think we've had maybe a little less fiscal discipline. So things that are not as high a priority are getting funded," he explained, adding that in the 1990s he had a lot more fiscal worries than today.

"Getting a plan has been key," he added. "Getting a plan that our stakeholders buy into. I mean, our Canadian space industry really put its shoulder behind it this time in everything from logging, to advertising, to you know, trying to talk it up."

He paused. "But that industry's a lot smaller than it was in my time." The CSA has had a flat budget for many years (save for stimulus funding), he also pointed out. It's too simplistic to say the two are related, but perhaps we can see the priorities for Canadian spending overall when considering these two phenomena.

It's useful here to take a step back and consider Canada's position as a space power. I've read through all of our country's space plans over the decades, and repeatedly we have taken the strategy of acting in concert with the United States. That country has always been more heavily invested in space policy, in the sense that they spent federal funds on spacecraft and rockets and launch operations — something that Canada never aspired to do in a large sense, although there have been calls to build a rocket launching facility on the East Coast over the years.

Even in the United States, however, space strategy only goes so far. Everyone remembers Kennedy's call to send humans to the moon in the early 1960s, but few link his policy announcement with the fact that the Cold War was at its height and much of the country — including the military-industrial complex — was focused on beating the Soviets. When the Bushes — that would be George H.W. Bush and George W. Bush — announced their own moon plans in the 1980s and 2000s, those quickly died due to a lack of interest.

Presidential initiative is only so powerful, as the famous 1997 space policy book *Spaceflight and the Myth of Presidential*

Leadership (by noted historians Roger D. Launius and Howard E. McCurdy) points out. In it, the authors argue that space policy is ancillary. The book is nearly 25 years old and hasn't been followed up in book-length form, although Launius pointed me to a series of monographs by W. Henry Lambright, a professor of public administration and political science at Syracuse University.

Some of these monographs read like a sequel to the 1997 book. Take the Bush Jr. situation, for example. The NASA administrator during Bush Jr.'s tenure was Michael Griffin, and "Griffin believed deeply in the moon to Mars mission," says one monograph. But despite a speech by Bush literally promising the moon, "the biggest obstacle [Griffin] faced throughout his tenure was the failure of the White House and Congress to fund the new mission and NASA adequately."[8] While a newly elected Obama endorsed returning to the moon by 2020 as Bush did, NASA later abandoned the idea as the program fell further and further behind due to funding woes. Once again, the president could only do so much against competing interests, even though he promised a moon mission.

The joke in Canada is if the United States sneezes, our country will catch the cold. Canada was the first country to sign on to the Gateway project outside of the United States, which was a visionary and bold thing to do even if the timing seemed questionable to some observers. More worrisome, however, was the United States changing tracks slightly just a few weeks after Canada's astronaut-studded announcement in 2019. US vice-president Mike Pence announced that his country would return astronauts to the moon's surface by 2024 — that would be the end of Trump's second term, if he is re-elected, and much faster than the previous pledge to send astronauts to a space station above the moon sometime in the 2020s.

To explain the issue: orbiting a spacecraft around a planet is in some ways easier than landing it, although this oversimplifies the elegant art of orbital mechanics. The Apollo 8 crew managed to make it all the way to the moon and back in just a single spacecraft in 1968 simply because they weren't landing. But a year later, the Apollo 11 crew needed two spacecraft because they were landing on the moon and then coming back through Earth's atmosphere.

Landing on the moon requires a lightweight, flexible craft. Returning to Earth requires a spacecraft with heavy shielding. Those are two spacecraft with opposite goals, and while there were engineering solutions that could have addressed both in the 1960s, NASA elected to use separate craft for simplicity and time reasons. Five decades later, we're still designing spacecraft in a similar way. So a moon lander would need to be a separate spacecraft from a moon orbiter. More spacecraft requires more money.

Moreover, landing on the moon is a different policy decision than orbiting the moon. In the wake of the announcement, some media and space observers worried that Canada's limited space dollars would face a competition between participating in the space station and participating in lunar landing missions.[9] Never mind that the US has never pressured Canada to do both, but simply asked for help with *something*. And while some were busy hand-wringing, the Canadian Space Agency did go ahead; by August 2019, it had awarded MDA a contract for Phase A (or preliminary) work on robotic interfaces for Canadarm3.[10] In other words, the work is moving forward — unless we hear otherwise.

Let's not forget another achievement of the Trudeau government that was lost on Canadian media: his government did release the long-awaited space strategy. The 22-page document[11] has a lot of fodder for the Canadian space industry; among its key tenets, those with the most direct economic benefits include creating an "AI-enabled deep space robotic system" (that would be Canadarm3), "guarantee[ing] the future of our astronaut program" (which, historically, has meant providing robotics to NASA in return for future flight opportunities), prioritizing Earth observation and collecting climate change data (which could mean more satellites) and providing support for international partnerships and space firms.

The lack of a space strategy was an issue that had dogged the previous Harper administration through nearly a decade of power. And Trudeau's government took nearly four years itself before releasing the strategy. The sense of relief must have been immense. In late 2018, a coalition of Canadian companies led by MDA got so frustrated with a decade of space strategy delays that it papered dozens of city buses in Ottawa with the slogan "Don't Let Go Canada." For a few incredible months, these vehicles paraded through downtown just five minutes' walk from Parliament Hill.

The frustration had come from the 2018 budget, which was supposed to prioritize space according to all the rumours that SpaceQ, a leading online news service covering the Canadian space sector (and a client of mine), had heard from Canadian industry. One of its articles outlined the state of affairs: "The space community is used to dealing with successive governments that don't prioritize them, however the 2018 budget was supposed to be different."[12]

The consultations and promises led to a budget below expectations, though, as the budget focused on pure science and some broadband satellites. In the months following the budget, according to CSA president Laporte at a space conference in mid-2019, the space community (industry and government together) "spoke with one voice" and also helped to change the perception that space only benefits a lucky few astronauts and engineers.[13] (Of course space technology benefits all of us every day — we can't use cell phones or watch TV or navigate cars or check the weather without it, something we easily forget.)

This is all very good news. Lack of commitment to space by previous administrations can be linked to issues such as MDA almost selling its Canadarm (and Radarsat satellite) technology, and the fact that no new Canadian astronauts were brought into the fold for an incredible 17 years — between 1992 and 2009. It is, of course, simplistic to think of governments as being space-ignorant or deliberately trying to belittle the industry as larger economic forces shape our budgets. Canadian prime ministers must work in the context of United States relations, military needs, economic booms and busts and (more recently) Indigenous relations when devising their priorities and policies. Taking a narrower focus on science, however, produces some insights.

Notably, 1992 and 2009 were both recession-heavy years, and it is probable the astronauts were brought on board as part of a larger effort to stimulate the economy with technology-rich investments. Canada was ruled by minority governments for much of the 2000s, making it difficult to push through many policies. And one must also remember that spaceflight opportunities for Canadians between 2003 and 2006 were delayed in the wake of Columbia's fatal incident, which required an investigation by NASA and numerous design changes and test flights before the shuttle was deemed safe to fly again.

Then again, we cannot forget that science policy by governments shapes what happens, at least to an extent. Let's take the most recent Harper and Trudeau governments as examples. University of Montreal political philosophy professor Christian Nadeau outlined a series of cuts the Harper government made in 2009 alone — such as removing a $325,000 subsidy to Mont-Mégantic Observatory and eliminating $150 million in funding to three major university funding bodies. "The least that can be said is that Gary Goodyear's tenure as science minister did not go unnoticed," Nadeau wrote in 2010.[14]

Harper's other notable science policy decisions included eliminating the long-form census, prioritizing science funding for projects that had direct relations to industry initiatives, shutting down the world-renowned Experimental Lakes Area and centralizing government communications to such an extent that scientists accused him of "systematic muzzling."[15][16] Or as one Canadian academic put more gently, "The new practice of having ministers' offices approve every press release and all public appearances for diplomats and scientists put a damper on the customary independence of these professionals."[17] However, Harper did include several space initiatives in stimulus funding in 2009 (such as creating new rovers for future lunar and Mars exploration), and his government approved the recruitment and hiring of two new Canadian astronauts that year. There was lots of bad news to write about, including a lack of long-term space strategy, but space did receive some benefits under his term.

The Trudeau government inherited a science community upset about Harper's initiatives and promptly reversed several of his policies — such as reinstituting the long-form census and relaxing the controls on government communications. His government created the Arctic and Northern Policy Framework, and as stated earlier, he also brought back Canada's national science advisor on the advice of the Senate. In space, his government committed to extending Canada's time on the International Space Station, recruited two astronauts and unveiled them to the public on Parliament Hill during Canada's 150th birthday, and made numerous smaller investments in artificial intelligence and satellites. While his space policy was only unveiled near the end of his first term (he was successfully re-elected to a minority government after a contentious 2019 campaign), Canadian industry does seem pleased with the move — at least to the extent that the buses with decals bemoaning a lack of space investment have vanished from downtown Ottawa streets.

But again, I thought the timing was interesting in regards to introducing the new space policy — while this was a clear win for the Canadian Space Agency and the space industry, all these announcements in an election year (and a year marred by the Wilson-Raybould scandal) was troubling. When the policy was announced in early March, Wilson-Raybould was still a constant in Canadian news coverage. This continued long after the national media moved on from space. Trudeau in fact expelled her from the Liberal caucus in early April (to the anger of opposition leaders).

Then in mid-August 2019, a report from Canada's ethics commissioner[18] showed that the rumours of the year were true — SNC-Lavalin had been lobbying for relaxed prosecution since at least 2016, and Trudeau had violated Section 9 of the country's Conflict of Interest Act. Trudeau responded by saying he took responsibility but did not agree with all of the findings. "Where I disagree with the commissioner is where he says that any contact with the attorney general [Wilson-Raybould] on this issue was improper," Trudeau said in a statement to media.[19] With Wilson-Raybould expected to release a biography in September 2019, more revelations will likely come in the next year or so.

I'm not a political reporter, and even when I was a journalism student at Carleton University, I turned away opportunities to go to Parliament Hill — considered a prime reporting location by many of my politically savvy classmates — because I was laser-focused on science coverage. My understanding today, with a little age, is more nuanced — that science and policy are intertwined, and both must be considered in writing stories. But my understanding comes from being an outsider and observer to political reporting. Those who follow the prime minister's dealings day to day may have a more informed stance.

However, I wanted to hear from leaders at the CSA to get their thoughts about all of these successive wins. From what I could see, they had been unexpectedly brought into a political maelstrom in 2019 when what they had been really trying to achieve for years was a consistent space policy for industry. So I spoke with some of the senior leaders at CSA to find out what the new announcements will mean, in the context of Canadian space leadership.

When CSA president Laporte and I spoke in May, Saint-Jacques was wrapping up the longest stay yet by a Canadian in space — more than six months. While his crew was limited to three people for much of the stay, they came close to meeting the original expectations for a crew of six, Laporte said. "The ISS partnership [members] have been impressed by the amount of work that was carried out by the three astronauts that were on the station," he told me.[20]

Canada's leadership not only showed on the ground but also in space, as Saint-Jacques took the first CSA spacewalk in a decade and also shepherded an (uncrewed) Crew Dragon on its first test flight, using Canadarm2. Having a Canadian on board is not only good for our scientists but also meets one of the CSA's mandates: to inspire young Canadians and to encourage them to stay in school — particularly in the disciplines of science, technology, engineering and math. Laporte said Saint-Jacques did dozens of discussions with schools in all regions of the country, including Canada's North where Indigenous populations are heavily based. He had spoken directly with more than 3,100 students at the time of my conversation with Laporte.

But with Saint-Jacques's mission ending, the conversation quickly shifted to the policy environment for space in Canada. Laporte sees it as a far more diverse sector than back in 2015, when Trudeau's government assumed power. Canada's extension of ISS operations to 2024 (four years longer than planned before) was announced in 2015. In 2017, funding for two science missions was announced — one for quantum encryption, and the other for radar observation around Mars. Broadband satellite funding was announced in 2018, allowing for faster communications across the country.

But what Laporte pointed to as well is investments outside of space that can still benefit space exploration, including the Strategic Innovation Fund, the Innovative Solutions Canada fund and the granting council Canada Foundation for Innovation. These are important, he said, because they allow CSA to encourage the space industry and research sectors to "tap into those opportunities for them to get funding." While he didn't have any firm numbers on hand, he said anecdotal accounts show him that "space has been quite successful in securing funding from those other sources."

"So now we've got an environment where the space program folks in Canada have access to a whole lot more different types of funding, and a whole lot more funding, than to rely solely on the budget of the CSA," he added, which is a noteworthy development in my eyes. CSA budgets tend to be fairly flat from year to year, so to move the space industry into other — related — sectors eases the pressure a bit on CSA projects. It also enhances the CSA's linkages with other government departments and industries — and he argues this makes the CSA competitive on the world stage.

"Since its inception and for decades now, the CSA has had that culture, that perspective, that we nourish industry and we nourish researchers in Canadian universities," Laporte said. "So when you look at what's happening in space today, with an increase in international missions, we're already there. When you look at commercial space and having an industry that competes globally to win contracts, well, we were already nourishing that for decades."

He argues this also positions the CSA well to work in the new space environment, which is filled to the brim with independent space companies competing for launch contracts — names such as Sierra Nevada, Blue Origin, Moon Express and Airbus all came to Laporte's lips. These are not only well-known names in the space industry, but they also have all expressed interest in "large Canadian space contracts," so the CSA has brought them for business-to-business events and shown them what our industry is made of. CSA has also participated in international summits to bring venture funding into our industry. This is all early-stage work, but Laporte said he is optimistic about the result.

But most noteworthy in his eyes is the $1.9 billion promised over 24 years for Gateway. That's a relatively small amount per year — roughly $80 million — but some of our contributions may be paid for through in-kind work, just like we do today. Additionally, Canadians have always worked with small space budgets, and as such, pursue projects that give a large return for a relatively small amount of money invested. The big question is whether this support will continue, but space always faces this challenge of attracting the attention of new administrations.

Laporte expressed optimism about the current space environment and said that since the accelerated timeframe became apparent, the CSA has worked with NASA, the European Space Agency and other international partners to start learning what this will mean. While the CSA is still at the options analysis stage, what he knew for sure was "NASA is looking at a very critical role for Canadarm3 in terms of assembling the Gateway and conducting operations out there in the proximity of the moon."

And while this book focuses on astronauts, Laporte said he wanted to mention some of the other name-brand projects CSA is working on today — the OSIRIS-REx mission to explore an asteroid, the new Radarsat Constellation satellites that enhance our ability to track climate change across the country, and the Canadian pointing device that will help the future James Webb Space Telescope look at distant objects. (Webb is far behind schedule and over budget and has been accused of cannibalizing NASA's astrophysics budget over the years, but at the time of writing, it is in assembly and expected to launch in 2021.)

In CSA's structure, the most influential person in human spaceflight — besides Laporte himself — is Gilles Leclerc, the director general of space exploration. He's been connected to the CSA almost continuously since 1989, save for a short break between 1997 and 2000 when he went to the Canadian embassy in Paris to be a delegate to the European Space Agency. During his current role, which started in 2010, the space station made the rapid shift from finishing construction to ramping up operations. There was international pressure to make the space station worth the money and effort that governments worldwide had poured into it.

As for any big science project, the results of the space station at this still early stage are hard to quantify. It may be that way for many years. Many of the commercial operations on the ISS are managed by the Center for the Advancement of Science in Space. It came under criticism in 2018 by NASA's Office of the Inspector General for "underperform[ing] on tasks important to achieving NASA's goal of building a commercial space economy in low Earth orbit." Yet the company can point to success in in-space manufacturing, including in the areas of 3D printing and pharmaceuticals, since assuming its role in 2011.[21]

For his part, Laporte is very focused on the space policy side of things — and he too can point to successes. While he did have some difficult moments — several Canadian astronauts retired within a few years as they aged out of the program — he has overseen new Canadian astronauts coming in their place and the extension of space station operations to at least 2024, which has helped smooth the discussions for Gateway and perhaps further extensions to ISS operations, Laporte said.

He called the CSA Gateway commitment crucial, since that was the direction the international committee was trending towards during our interview in March (prior to NASA's moon-landing announcement), although how the participation will happen is still being defined. It may be that Canada simply extends the existing intergovernmental agreement to continue with Gateway. Or there may be some changes to reflect a service-type agreement whereby Canada and Canadian industry provide services.

I asked Leclerc about the timing of the Canadarm3 and space strategy announcement, adding that Ministers Navdeep Bains and Garneau both defended these occurring in the fourth year of their mandate; negotiations, especially when it comes to cutting-edge technology, take a lot of time.

Leclerc joked that he's "not going to comment on two ministers" — a fair comment, especially since Bains is his boss and Garneau's word still carries a lot of weight in the space community. But what he did say was that in 2018–19, "the stars were perfectly aligned." A lot of good things came together, including the new funding commitments and the astronaut recruitment campaign.

"I mean, given the fact that this was the fourth year of the mandate is a nice coincidence, but one that had to happen at some point, because there is pressure on all the partners now to commit to Gateway, to the lunar program. And we're extremely proud that we're the first ones to engage with NASA," Leclerc pointed out.

It's possible that an observer reading this book in 20 or 30 years may wonder why there was so much attention paid to the Wilson-Raybould situation when the results of the space program probably will speak for themselves: Canadians working at or near the moon and an active Canadarm3, if the dreams of today come to pass. Making firm predictions about space is a fool's game, but from the perspective of 2020, it is very difficult for any futurologist to talk about what will be possible in a generation.

That said, the message to take away is that while space policy works hard to be apolitical — in the sense that partners work together even amid such international crises as the Russia–Ukraine controversy — in some cases the timing of funding announcements, or the funding itself, can be linked to political aims. The true test of the 2019 announcements will be whether they are sustainable enough for the United States and Canada to continue working together on these projects, or whether (as both countries go through elections) they will dissolve into thin air. And that's where our legacy, our position on the value of space, will really be important.

EPILOGUE

And what you thought you came for
Is only a shell, a husk of meaning
From which the purpose breaks only when it is fulfilled
If at all. Either you had no purpose
Or the purpose is beyond the end you figured
And is altered in fulfilment.

— T.S. Eliot, "Little Gidding" (1942)

"Sending astronauts into classrooms just isn't working."

I was at an event completely unrelated to space exploration — an Ottawa panel discussing female entrepreneurship, in June 2019 — when an academic said this and got a lot of applause for it. In a sense, she was right. The numbers of women in science, technology, engineering and math (STEM) in Canada remain abysmal after more than 30 years of effort. The statistics cannot be denied. In fact, there is much we need to do to open up these fields to people of *all* genders.

Luckily, there are many entities out there working to inspire marginalized people, including LGBTQ+ as of recently, to join the space program. (At this time, we know of only two LGBTQ+

people who have flown in space — Ride, whose personal life was only revealed to the world after her death and Saint-Jacques's crewmate McClain, who was outed in 2019 during a dispute made public by her estranged spouse.[1]) Out Astronaut is a collaboration of several organizations that aims to train more LGBTQ+ astronauts and eventually to fly a scientist-astronaut in space. I firmly believe diversity is a great thing in science, and I hope that more science-focused initiatives like this spring up in the future.

So I fully agree the STEM issue needs to be addressed, not only in terms of inspiring people but also in terms of blasting away barriers keeping people back. But at the same time, I know not everyone chooses a STEM career. I think back to all of the space-oriented people I have met who work in fields that are not "hard sciences." In the year that I wrote this book, two elementary schoolteachers asked me to speak to children (Grades 1, 3 and 4) about space — even though it wasn't in the curriculum. There were all the folks I encountered in Kazakhstan: the journalists, the communicators, the interpreters, the tour guides. These people won't show up in statistics about STEM. Yet they are not only inspired by space. They live space.

And there's Linda Dao. In December 2017, while covering preparations of the Saint-Jacques mission for Space.com, a daily news site out of New York City where I have been freelancing for almost a decade, I ran into Dao at an event at the Canadian Space Agency. She had won a nationwide contest in 2013 during the Hadfield space mission. In between undergraduate studies at McMaster University, she designed an experiment to demonstrate how surface tension between molecules works in space. It was a last-minute submission after she felt a previous one wasn't enough to win, she said.

Four years later — and four years is a long time for a person just entering the workforce — Dao remained passionate about space. She took graduate studies at the International Space

University and was a member of Students for the Exploration and Development of Space. She happened to be on site for a professional conference, and by luck, attendees were encouraged to attend Saint-Jacques's presser. I sat beside her, and we chatted for a long time, mostly for a story I was working on. Eighteen months later, on the eve of Saint-Jacques's landing, I looked her up again. This former winner of a Canadian Space Agency contest, as of that time I looked her up, *works* at the agency as a research analyst.

Then there was the manager at my bank. I don't spout off about space in all my business dealings, tempting as it is. But this manager (who I'll call Alina, to preserve her privacy) called me one afternoon asking me about the Russian ruble order I had made in advance of my trip to see Saint-Jacques's launch. The order she had in hand only asked for about $60 in rubles, and she correctly suspected there was a math error. Once we sorted out the money, she asked me why I was going — visits to Russia aren't so usual in Ottawa, I suppose. So I told her the story: I was a journalist, I was going to see a Canadian astronaut launch from Kazakhstan, it was something I'd wanted to do for a long time.

To my surprise and delight, when I came to the bank days later to pick up my rubles, Alina was there — and peppered me with questions about the mission. She had immigrated to Canada some years ago and had never followed space or thought much about astronauts before I mentioned my trip. But after our conversation, she looked up the Canadian Space Agency YouTube channel, and, watching Saint-Jacques speak, was very inspired by his positive message. And as the mission progressed, every time I visited the bank she would bring up Saint-Jacques. She even contacted me just before he landed, saying she was hoping for a safe return.

I can relate to these women's experiences of being struck by and inspired by space. In 1992, I met an astronaut (Bondar)

— not in a classroom but at a local museum. It wasn't until 1996, however, that a teenaged version of myself watched the movie *Apollo 13* and dreamed about becoming an astronaut. I read every book in the library about Apollo. Learning about space inspired me to learn more about the related sciences in school — biology, chemistry, physics.

This *almost* made me go into STEM. I considered aerospace engineering all the way up to the last year of high school, actually, but two things pushed me towards journalism — an editor at *Nepean This Week* telling me I was very good at it, and looking at my high school marks and realizing not only did I get better marks in English but I enjoyed it much more. And eventually, after many career twists and turns amid an economic recession in 2009, I ended up as a space journalist. You won't see my career show up in any STEM surveys, but I have been inspired by space for almost 25 years — and I expect that will continue for the rest of my life.

We can't all be astronauts, but we can use space as a launching point for other interests. And while space travel didn't work out for me (at least so far), I can name one person who was directly inspired by Bondar and who is on her way there. That's Sidey-Gibbons, Canadian astronaut and an eloquent spokesperson for the power of science, even though she's only in her early thirties.

"Roberta, I want to show you something I made 27 years ago, when you flew on the space shuttle," said Sidey-Gibbons in a 2019 Canadian Space Agency video.[2] The two astronauts sit side by side on armchairs as Sidey-Gibbons leafs through a scrapbook. "I think mostly my mom was the one who put the scrapbook together," the young astronaut admitted with a laugh. "She probably didn't trust you with the glue," Bondar responded — after all, Sidey-Gibbons was only three.

The conversation turned serious. "I wanted to show you this because my mom kept it all these years," Sidey-Gibbons added. The camera focused on a yellowing newspaper photo of Bondar, wearing her 1992 flight suit.

"I think parents have a tremendous role to play in our lives," Bondar answered. "And if I may, I just want to tell you that when I was a little girl, I used to put all these plastic model rockets together and then — when I got into the space program, the first Christmas there was a big box under the tree, with a nice red bow on it. And when I opened it up, inside were all those space models."

Bondar was a neurologist. When Sidey-Gibbons grew old enough to choose a career, she didn't become a doctor. Perhaps you could see that as a failure of some sort and say, look, "sending doctors into classrooms is not working" in terms of bringing more doctors into the Canadian medical system. But the simple answer to why Sidey-Gibbons looked elsewhere? She loved fire.

"Fire is something that we've used for hundreds of thousands of years, and we still didn't know everything there was to know about it," Sidey-Gibbons told me. "It was so fundamental to us as a species, and we were still studying it. There were a lot of cool concepts we didn't understand yet."

In particular, Sidey-Gibbons enjoyed learning about the turbulence of flames — the unsteady fluctuations in a fire. This turbulence can scale very quickly in a way that is not easily explained by mathematics — not doubling in complexity, for example, but becoming more complex so quickly that math cannot show us the relationship easily. Campers know how a small match spark quickly consumes the entire log. "You're dealing with so much at once that even a flame that's very small is changing so fast [that] it's very difficult to study," Sidey-Gibbons said.

Despite being in the astronaut program for a few years, Sidey-Gibbons keeps up in her original field of study and says that what is needed is more computing power. Canada is known for its speedy quantum computing possibilities, so the solution could very well come from this country — hopefully with people from a variety of backgrounds and genders tackling it together.

On June 24, 2019, I slipped on a pair of astronaut socks. These socks have special meaning to me. I received them as a gift in 2015 and have put them on during many special space moments: when Saint-Jacques was named to the crew, the day he launched in Kazakhstan, the day he stepped outside as a spacewalker. On this day, he was coming back to Earth.

I did have a nightmare related to his return — a vague dream about astronauts I had interviewed getting upset about bad news. Landing is also a dangerous activity, and I vividly remember the horrible day of Feb. 1, 2003, when Columbia broke up in the atmosphere on what was otherwise a beautiful Saturday morning in Texas. I watched the full coverage on television. To this day, I still get the shakes if I have to write an article about it. But I write, because others must remember.

My mother texted me on Saint-Jacques's landing day, telling me there was a small contingent of Canadians that journeyed to Kazakhstan. I wasn't interested in going after two expensive trips I self-funded there in 2018, and in an unusual move, I had even turned down the opportunity to watch the late-night landing from the CSA headquarters near Montreal. Sadly, an urgent house maintenance appointment kept me away.

But I did keep an eye on things. I had little access to Twitter while in Kazakhstan, but at home I could easily check in on how the crew was doing, minute by minute. Which I did, nervously. I trusted that the crew and ground team knew how to get everyone safely home, and that the spacecraft was built well, but you never know for sure.

Astronauts Thirsk and Hansen were on hand at the CSA headquarters that night, talking about the mission to the employees and journalists attending. There were some interesting moments in the broadcast, as often happens with "lives."[3] Broadcasting from Kazakhstan is no small feat as cell service is scattered — but the Russians are excellent engineers and know exactly which satellites to tune into, letting us "experience" being on the steppe right alongside the crew.

The astronauts landed safely (cue the sigh of relief) and, as usual, were carefully carried out of their little capsule to lawn chairs on the sunny Kazakh steppe. Watching Saint-Jacques surrounded by doctors and Russian officials, my mind flashed back to seeing him striding tall and triumphant in that region just a year before — positively glowing as a member of the backup crew, and just six months away from going to space.

Now, though, he looked ill — a common issue for astronauts adjusting to gravity after six months aloft. The camera spent little time on him to preserve his privacy, but CBC (who was on site at the landing) reported he was "extremely nauseous." Astronauts are very commonly sick upon landing, if not on camera then inside the medical tent where they practise walking again in private. But doctors are used to helping the astronauts adapt to gravity. Within a few weeks, most astronauts can drive again and stand up in the shower.

"Gravity is not my friend," Saint-Jacques told CBC a few hours later, outside the medical tent and propped up in a standing position by two medical personnel. Yet he still found some pleasure amid the nausea. "What's striking me is the grass, the smell of the grass here. It's just beautiful."[4]

Saint-Jacques's crewmates had more time on screen during the NASA broadcast. McClain found the energy to put on her own ball cap and playfully point to the camera, smiling exuberantly

— while being careful not to move too quickly. Kononenko laid back in his chair and looked serene.

Then people began handing out apples. Rookie McClain gently held out her hand and smiled politely in a gesture of refusal, and the crowd at CSA (captured on video) began laughing in sympathy. Kononenko, however, was on his fourth trip home from the space station and knew exactly how to hold his neck and head to minimize wooziness. So he didn't blink when offered — he reached for the apple. Took a bite. Chewed, carefully showing no reaction on camera. Leaned his head forward and bit again.

"He's showing off," Thirsk joked as the crowd roared.

"He *really* wants to fly again," Hansen added, then mused on Kononenko's thoughts: "'I'm going to eat this apple.'"

Left unsaid by the astronaut commentators was if and when Saint-Jacques would fly again. After waiting nine years for a spaceflight, it seems more than likely that Saint-Jacques will step aside to let the three other astronauts — all unflown — have their chance in space. Through Thirsk, I detected some wistfulness from the retired astronaut when he spoke about an email that Saint-Jacques sent him a few days before returning home.

"He described the six months as being surreal. Magical," Thirsk said, using the words we want to hear from astronauts — the required gushing about Earth's view, about working in microgravity, about being heroes for a few months. But Thirsk then lapsed back into how astronauts actually measure their work efficacy on station — how much their work efficiency increases as they get better navigating in weightlessness. In an environment where the crew runs 200 experiments and has days scheduled by five-minute increments, little changes matter.

Most astronauts awkwardly use too many limbs to push around in microgravity during their first couple of weeks, but after a time, they develop a unique and elegant form of "swimming" — they use their arms to pull from location to location and allow their legs to trail behind, only lifting or lowering those lower limbs to go around obstacles. The astronauts get used to working while standing on their heads, or inside tight corners.

"In one respect, he was glad to be coming home, to reunite with his family," Thirsk added about the Saint-Jacques email. "But in the other respect, he was a little bit sad because he's finding that his ability to perform as an astronaut — his comfort with space-life, his work productivity — was surprisingly still increasing after already having spent almost seven months in space. And he wanted to see what his limits were. So that was an interesting perspective. For astronauts, it's intriguing to not only explore new territories but also to explore the limits of our own mental, physical and emotional beings."

Thirsk, retired as an astronaut for almost a decade, paused before continuing. "I understood what David was saying," he said with a sad smile. "I identified with the emotional dilemma he was experiencing."

When I began writing this book, the conversation about Canada in space was a much different one. As I signed my contract in October 2018, city buses in Ottawa — which trundle through a bus corridor less than a kilometre from Parliament Hill — had that space-themed wraparound: "Don't Let Go Canada." I had never seen such a strong statement about space in Canada before or since.

Those star-wrapped buses were a protest by a coalition of Canadian aerospace companies (including big names such as MDA) concerned about Canada's lack of a long-term space plan. Canada didn't have a space plan when I was a young reporter in 2009, watching astronauts Hansen and Saint-Jacques get introduced to media for the first time in Ottawa. Canada still didn't have a space plan after Hadfield became world-famous in 2012. And in late 2018, after the NASA president personally visited Ottawa and announced at a prominent space conference he needed Canada to agree to a moon mission soon, no plan appeared forthcoming still. Companies were restless. They wanted answers.

Everything changed after Saint-Jacques went into space. With the annual federal budget looming, in the year of a federal election, and as the Trudeau government grappled with a sponsorship scandal, suddenly space was in the national conversation for the first time in half a decade. In swift succession, the CSA and several ministers announced a Canadarm3, the opportunity to bring Canada to the moon and a space plan emphasizing the roadmap to get there. Meanwhile, NASA accelerated its moon plans to land astronauts by 2024 (near the end of Trump's second term, if he is re-elected, or within the first term of a new president's mandate).

I can't tell you if Gateway will be built on schedule or if astronauts will walk on the moon in less than half a decade, but it's common for space programs to be pushed back. For example, NASA was hoping to have commercial crew vehicles ready almost immediately after the shuttle retired in 2011; eight years later, Saint-Jacques finally saw the first uncrewed version arrive at the ISS. (He was in fact the first-ever astronaut to go inside, which is a Canadian milestone that most history books will probably ignore.) And in the pages of this book, you have seen the stumbles on the way to building the International Space Station, the false moves and programs that were supposed to bring us beyond Earth but never came to pass.

So what is next for the ISS and its astronauts? As I edit this epilogue in May 2020, many space industry events and activities are being disrupted by the rise of the novel coronavirus pandemic, whose long-term consequences cannot be predicted at the time of this writing. That said, the next initiative is

Commercial Crew — those company-led spacecraft that SpaceX and Boeing have been working on for the better part of a decade. The first crews have been announced for the test flights, naturally composed of all American astronauts due to the fact that it's a NASA program — and probably the inherent risk of being among the first crews to fly it.

Once those test flights are through, NASA expects the flight pace to pick up — and that Canadians will be able to fly more often, too. ISS manager Shireman told me so in March 2019, adding, "Hold me to it. Come back and talk to me in another year and a half and see if I'm true to my word."[5] I suppose that's because at that point NASA expects to be finished with the first test flights and to announce newer flights for Commercial Crew.

While we eagerly await a bigger Canadian manifest, it is interesting to note a surprising Canadian connection with one of the first Commercial Crew members.

Michael (Mike) Hopkins was selected in the same class as Hansen and Saint-Jacques, in 2009. While he's not Canadian, he spent years living in Canada on an exchange program. This former US Air Force test pilot moved from sunny California to frigid Cold Lake, AB, in 1999. He thrived in the small community of 12,000 people, staying so long that his second son was born there and earned dual citizenship.

Hopkins, meanwhile, flew a who's who of Canadian aircraft platforms — the F-18, the C-130, the CH-146 helicopter, the Tudor and the C-33. "Just had an incredible experience professionally, but as well on a personal front," he said.[6] He also remembers barbeques in −20 degrees: "You accept it and you embrace it and you continue to enjoy the lifestyle, the outdoors and all of that, even though [it's] in the middle of winter. Now when it got minus 40, you were hunkering down, but that happened rarely."

Perhaps that experience helped for his 2009 astronaut application, or for the six months Hopkins has already spent in space. He's now getting ready for Crew Dragon. The stay aboard the space station is going to be the same as usual, but where Hopkins and his crewmates — NASA astronaut Victor Glover, NASA astronaut Shannon Walker and Japanese astronaut Soichi Noguchi[7] — get to make a difference is in the spacecraft training. They'll be helping to write the procedures for future crews, and they'll be pointing out how these vehicles are similar or different to the Soyuz that most astronauts today are familiar with.

There's another thing Hopkins is looking forward to. Crew Dragon can hold four people. Soyuz can only hold three. And this means, by pure numbers, there will be more astronauts going into space. He praised the Canadian corps — most especially Hansen — for their patience in waiting in line and added that he hoped the situation would change when Dragon is available.

"The Canadian astronauts unfortunately don't get a lot of seats," he said. "They don't fly that often. I don't know the program's plans for how they're going to fill the seats that we have, but we have an extra seat. We have four seats versus just three that the Soyuz had. That, I hope, is potentially going to make differences in terms of the fly rates, not just for US astronauts and European astronauts, but also the Canadian astronauts."

So for the foreseeable future, Canadians will still be going to ISS — although the Trump administration, as of May 2020, did hope to land astronauts on the moon by 2024. It's unclear if NASA can meet that deadline as it will require tens of billions of dollars, not to mention marshalling the resources of its thousands of workers — this amid and after a global pandemic that is depressing economies worldwide. Canada might end up going to the moon, too, but for now, we know at least we'll return to ISS.

We're lucky ISS exists. Against all odds, a fatal shuttle flight and a Soyuz abort and all the political issues over the years, it survived. And it still can inspire at odd moments.

Most of the world's population can see ISS pass overhead, a steady star that skims across the sky for as long as six minutes at a time. I remember my journalist friend Sean Costello, near the end of our otherwise difficult trip to Kazakhstan in December 2018, encouraging all of us to go outside to "wave" to Saint-Jacques in his Soyuz as he caught up to the ISS, just hours after his rocket launched in front of us.

It was cold, it was dark, we were jetlagged and tired from a long few days, and we were in a strange (to us) country surrounded by little more than barbed wire and desert. But looking up at the stars, watching the faint Soyuz passing overhead, I felt at home again.

There was a Canadian making his way in an even more hostile environment than Kazakhstan's December cold. There was a Canadian thriving amid the stress and fame of a spaceflight. There was a Canadian whom history will remember, long after all of us now living turn into dust.

There was a Canadian who flew among the stars.

Who will be next?

AUTHOR'S NOTE

Many technical histories have been written about the Canadian space program. Instead of duplicating their excellent work, I took a different tack and tried to write a book that a non-specialist would enjoy. My goal was a broad overview of Canadian space work from the 1960s to today, focusing on the narrative of the astronaut program and how it has matured through the International Space Station. The idea was to attract the attention of the public and politicians and help readers understand the importance of the space program in everyday life.

This book is the culmination of my 25 years of interest in the Canadian space program, and represents about 18 months of hard work by myself and a team of able helpers (most notably at ECW Press, to whom I am grateful for this opportunity). I used many sources in gathering materials: personal interviews with many people in the space program, including most of the astronaut corps; reading presidential and prime minister autobiographies and biographies; the existing Canadian space literature; and my own experiences from watching David Saint-Jacques launch from Kazakhstan in December 2018. Where I could, I included translated information from Russian; despite five years learning the language, my skills remain too poor to attempt much reading beyond elementary schoolbooks.

My interview sources fact-checked my words, and many editors went through my work, but any mistakes that remain are my responsibility alone.

ACKNOWLEDGEMENTS

I am grateful to the Canadian Space Agency (CSA) and its network of partners (including NASA, the European Space Agency, and Roscosmos) that helped me and other Canadian journalists view the backup and prime launches of David Saint-Jacques in June and December 2018. The CSA was an essential partner in my book's logistics. The team granted me research access at their headquarters, interviews with their active astronauts and senior personnel, and fact-checking besides.

Thank you to every person who agreed to be interviewed, including their many support staff who made this a priority among otherwise very crowded schedules. Special thanks to astronaut Chris Hadfield, whose kind recommendation of my work helped launch the book deal, and astronaut Dave Williams, who voluntarily fact-checked the whole manuscript and provided a foreword showing his unique perspective on Canadian space history as an explorer of space himself. I am also in debt to the personnel working in early 2019 at the offices of the Governor General (Her Excellency Julie Payette, an astronaut), the minister of transport (Marc Garneau, astronaut and former CSA president), and the deputy minister of Veterans Affairs (Walter Natynczyk, former CSA president). It was a humbling and rewarding experience to speak with these busy people, who help run the affairs of Canada as their day job.

This book is principally before you thanks to the confidence of two men — my editor (and long-time mentor), Tim Lougheed, and my publisher, Jack David. Tim pitched me as a Canadian space program expert and Jack believed him, even though I was only in my mid-30s when I signed the deal. I've always wanted to write this book, and I cannot believe it happened so soon. To every other person at ECW Press and its network of partners who shepherded this to publication — I am grateful.

Numerous writing mentors influenced me over the years. At a distance, the entire movie team of *Apollo 13* introduced teenaged me to the excitement of solving problems in the space program; thanks to director Ron Howard, lead actor Tom Hanks, and book co-authors Jim Lovell (who flew on Apollo 13 in 1970) and Jeffrey Kluger (space journalist) for bringing an incredible real-life space rescue to publication. I still have two of your movie posters on my office wall, reminding me why it's important to write about space for the "next generation" of readers.

Derek Dunn and Tim MacLean (the latter being the brother of astronaut Steve MacLean) at the now-defunct *Nepean This Week* convinced a space-crazed teenager and writer to take journalism at Carleton University. Science journalism professor Kathryn O'Hara wrangled $600 from Carleton for me to fly to St. John's; there, I joined the predecessor organization to Science Writers and Communicators of Canada and got a much-needed career boost. Carleton adjunct professor Peter Calamai (who sadly died suddenly and shortly after my book deal) kept bringing me journalistic opportunities long after I graduated. I wish I could have thanked him better and interviewed him about seeing the Apollo 11 astronauts visit Ottawa.

Thank you to the editors and news outlets that had me on staff in my early journalism career: the *Charlatan at Carleton University*, the *Globe and Mail*, the *Canadian Medical Association Journal*, the *Ottawa Business Journal* and CTV/CJOH Ottawa. Thank you also to the small but mighty Space Studies department staff at the University of North Dakota. They supported me as I worked part-time and at a distance for 10 years through an M.Sc. (under David Whalen) and a Ph.D. (under James Casler). In 2020, I also taught a course at UND for the first time; I am ever so grateful to that community, as well as all the other post-secondary institutions in Ottawa who have accepted me for contracts as a part-time communications instructor.

As a freelancer, you must live to serve your clients — and fortunately, I have generous ones. Thank you to everyone who has ever hired me, especially Marc Boucher at SpaceQ (who helped me attend space shuttle launches STS-129, STS-130 and STS-131 in 2009-10) and Tariq Malik at Space.com (who supported a two-week journey to the Mars Desert Research Station in 2014, and several conferences). Both SpaceQ and Space.com were my chief financial supports for the two Baikonur trips.

Even a solopreneur needs business support when writing a book. Virtual assistants Marc and Sue Morin, and my background editor and writer Christina Goodvin, kept my business running. Numerous friends and fellow entrepreneurs kept me and my motivation on track. Several workers at Rev.com transcribed my interviews while holding to a confidentiality agreement, saving me many hours of time.

A huge network of Canadian and American friends helped me through book writing. Special shout-out to my Baikonur travel partner and space photographer Sean Costello (who looked at an early copy of this book); book author Chris Gainor (who also reviewed the manuscript and provided transcripts of people I cannot today interview); and book author Jonathan Rotondo (who gave me valuable advice as I considered the book deal).

Personnel at the following organizations lent me books or research materials and/or gave me ideas for interview subjects:

the Ottawa Public Library, Carleton University, the University of North Dakota, the National Research Council, Library and Archives Canada, Ingenium Corp., and the Friends of the Communications Research Centre. Wally Cherwinski, formerly of the National Research Council, lent me a precious trove of magazines and space agency publications dating back to the early 1980s, which helped me better understand the excitement of the early astronaut program days.

Last but not least, my family. My parents introduced me to aerospace as an infant when they brought me to the Enterprise space shuttle landing in Ottawa in June 1983. My dad began his career as an airplane maintenance engineer, and as a child, I got to walk into the prime minister's airplane, fly to Michigan in a Douglas DC-3 and sit in the co-pilot's seat on a Cessna as my dad flew around Ottawa. I didn't know how cool that was until years later — sorry, Dad. My mom willingly pulled me and my brother out of school to see Roberta Bondar visit the Canada Science and Technology Museum in 1992, even though she was so pregnant she could barely stand. My sister wasn't born back then, but both she and my brother were unquestionably there ever since, whenever I needed research help or advice.

I knew I would marry my husband, J, on Date 3 when I mentioned New Mexico's Very Large Array by acronym only. He not only knew the acronym, but said he wanted to visit there with me. (We did.) J fed our cat Gabriel, cleaned the house and shovelled Ottawa's endless snow out of our yard while I was getting all the fun and attention in Russia and Kazakhstan. His generosity knows no bounds. I cannot thank him enough for his support and love.

— ELIZABETH HOWELL, MAY 2020, OTTAWA

ENDNOTES

Prologue

1   NASA. (2018, Oct. 11). *Crew safe after Soyuz launch abort* [Video]. YouTube. Retrieved from https://www.youtube.com/watch?v=LUwnLFKfuBE.
2   NASA. (2018, Oct. 16). *Q&A with astronaut Nick Hague on launch anomaly and safe landing* [Video]. YouTube. Retrieved from https://www.youtube.com/watch?v=vqEpCDCnduU.
3   Rowe, C. (2019, April 4). Personal interview.
4   Sidey-Gibbons, J. (2019, March 1). Personal interview.
5   Ruptly. (2018, Dec. 2). *LIVE: ISS Expedition 58/59 hold pre-flight presser in Baikonur* [Video]. YouTube. Retrieved from https://www.youtube.com/watch?v=TkPjQB7X8Ow.

Chapter One

1   Fairbairn, J. (1969, Dec. 5). Armstrong gets prairie relic. *Winnipeg Free Press.* Retrieved from the NRC Archive in Ottawa 2018, Dec. 1.
2   The Canadian Press. (1969, Nov. 29). Moonmen facing busy schedule. *The Ottawa Citizen.* Retrieved from the NRC Archive in Ottawa 2018, Dec. 1.
3   UPI. (1969, Nov. 18). Lunautes à Montréal le 3 décembre. *Le Journal de Montréal.* Retrieved from the NRC Archive in Ottawa 2018, Dec. 1.
4   Heward, B. (1969, Dec. 3). Children crash barricades to see moon men. *The Ottawa Citizen.* Retrieved from the NRC Archive in Ottawa 2018, Dec. 1.

5   Calamai, P. (1969, Dec. 3). Visit of astronauts overwhelms Ottawa. *Edmonton Journal*. Retrieved from the NRC Archive in Ottawa 2018, Dec. 1.

6   Astronauts predict crews from many nations. (1969, Dec. 3). *The Globe and Mail*. Retrieved from the NRC Archive in Ottawa 2018, Dec. 1.

7   The Canadian Press. (1969, Nov. 27). Quiet welcome in Montreal awaits Apollo 11 astronauts. *Prince George Citizen*. Retrieved from the NRC Archive in Ottawa 2018, Dec. 1.

8   Office of the Prime Minister. (1969, Dec. 9). SL28-69. Re: Astronauts' visit - complaint that public left out. Retrieved from the NRC Archive in Ottawa 2018, Dec. 1.

9   Calamai, P. (1969, Dec. 8). Canada's space program seems grounded. *The Edmonton Journal*. Retrieved from the NRC Archive in Ottawa 2018, Dec. 1.

10 Calamai, P. (1969, Dec. 8). Canada's space program seems grounded. *The Edmonton Journal*. Retrieved from the NRC Archive in Ottawa 2018, Dec. 1.

11 UPI. (1969, Nov. 26). Astronauts visit Ottawa Tuesday. *Wallaceburg Daily News*. Retrieved from the NRC Archive in Ottawa 2018, Dec. 1.

12 Astronauts predict crews from many nations. (1969, Dec. 3). *The Globe and Mail*. Retrieved from the NRC Archive in Ottawa 2018, Dec. 1.

13 Lindberg, G. (2019, Feb. 6). Personal interview.

14 Franklin, C. (2019, March 7). Personal interview.

15 Canadian Space Agency. (2018, Sept. 28). Alouette I and II. Retrieved from http://www.asc-csa.gc.ca/eng/satellites/alouette.asp.

16 Kirton, J. (1985). *Canada, the United States, and Space.* Toronto: Canadian Institute of International Affairs. p. 10.

17 Bourgeois-Doyle, D. (2004). *George J. Klein: The Great Inventor.* Ottawa: NRC Research Press.

18 Special thanks to the museum's David Pantalony for leading the tour.

19 Northrop Grumman. (2019). Storable Tubular Extendable Member. Retrieved from http://www.northropgrumman.com/BusinessVentures/AstroAerospace/Products/Pages/STEM.aspx.

20 Kirton, J. (1985). *Canada, the United States, and space.* Toronto: Canadian Institute of International Affairs. p. 10.

21 Bourgeois-Doyle, D. (2004). *George J. Klein: The Great Inventor.* Ottawa: NRC Research Press.

22 Jelly, D.H. (1988). *Canada: 25 Years in Space.* Montreal: Polyscience Publications.

23 Jelly, D.H. (1988). *Canada: 25 Years in Space.* Montreal: Polyscience Publications.

24 Lindberg, G. (2019, Feb. 6). Personal interview.

25 Lindberg, G. (2019, Feb. 6). Personal interview.

26 Truly, R. (2019, Jan. 14). Personal interview.

27 Lindberg, G. (2019, Feb. 6). Personal interview.

28 Ower, C. (2019, April 17). Personal interview.

29 Hiltz, M. (2019, April 17). Personal interview.

30 University of Calgary. (2008). NeuroArm procedure a world first. Retrieved from https://www.ucalgary.ca/about/our-story/our-history/storylines/human-and-animal-health/neuroarm-procedure.

31 Walton, D. (2007, April 18). As a matter of fact, it is rocket science. *The Globe and Mail.* Retrieved from https://www.theglobeandmail.com/life/as-a-matter-of-fact-it-is-rocket-science/article683553/.

32 Lindberg, G. (2019, Feb. 6). Personal interview.

33 Marsh, J.H. (2012, Feb. 19; edited 2015, March 4). The Avro Arrow is cancelled. In *The Canadian Encyclopedia.* Retrieved from https://www.thecanadianencyclopedia.ca/en/article/avro-iarrowi-there-never-was-an-iarrowi-feature.

34 Gainor, C. (2001). *Arrows to the Moon.* Burlington, ON: Apogee Books. p. 31.

35 Marsh, J.H. (2012, Feb. 19; edited 2015, March 4). The Avro Arrow is cancelled. In *The Canadian Encyclopedia.* Retrieved from https://www.thecanadianencyclopedia.ca/en/article/avro-iarrowi-there-never-was-an-iarrowi-feature.

36 Aikenhead, B. (1995, Sept. 25). Interview with C. Gainor.

37 Aikenhead, B. (1995, Sept. 25). Interview with C. Gainor.

38 Gainor, C. (2001). *Arrows to the Moon.* Burlington, ON: Apogee Books. p. 268.

39 Siddiqi, A. (2019, July). Why the Soviets lost. *Air & Space Magazine.* pp. 30–35.

40 National Air and Space Museum. (n.d.). Soviets lead the race. *The Moon Decision.* Retrieved from https://airandspace.si.edu/exhibitions/apollo-to-the-moon/online/racing-to-space/moon-decision.cfm.

41 Siddiqi, A. (2019, July). Why the Soviets lost. *Air & Space Magazine.* pp. 30–35.

42 Siddiqi, A. (2019, July). Why the Soviets lost. *Air & Space Magazine.* pp. 30–35.

43 Zak, A. Largest explosion in space history rocks Tyuratam. *Russian Space Web.* Retrieved from http://www.russianspaceweb.com/n1_5l.html.

44 Godefroy, A. (2017). *The Canadian space program: From Black Brant to the International Space Station.* New York City: Springer. pp. 73–77.

45 Chapman, J.H., Forsyth, P.A., Lapp, P.A., & Patterson, G.N. (1967). *Upper atmosphere and space programs in Canada.* Science Secretariat, Government of Canada, Catalog No. 582 -7/1967. Ottawa: Queen's Printer and Controller of Stationery. p. 3.

46 Godefroy, A. (2017). *The Canadian Space Program: From Black Brant to the International Space Station.* New York City: Springer. pp. 73–77.

47 Allaway, H. (1979). *The Space Shuttle at Work*. Washington, DC: NASA: Scientific and Technical Information Branch, and Division of Public Affairs.

Chapter Two

1 Garneau, M. (2019, March 15). Personal interview. [Unless otherwise noted, all quotes from Garneau in this chapter come from that interview.]

2 Côté, F., & Bonikowsky, L.N. (2011, Oct. 16; edited 2016, Feb. 16). Marc Garneau. In *The Canadian Encyclopedia*. Retrieved from https://www.thecanadianencyclopedia.ca/en/article/marc-garneau.

3 Garneau keeps seat in NDG-Westmount. (2015, Oct. 20). *CTV Montreal*. Retrieved from https://montreal.ctvnews.ca/garneau-keeps-seat-in-ndg-westmount-1.2617984.

4 Associated Press. (2005, June 1). Astronaut's hair sparks legal hubbub. *NBC News*. Retrieved from http://www.nbcnews.com/id/8062442/ns/technology_and_science-space/t/astronauts-hair-sparks-legal-hubbub/#.XzPpTy2z3UJ.

5 Cherwinski, W. (2019, Feb. 22). Personal interview.

6 Memorandum of Understanding Between the National Aeronautics and Space Administration and the National Research of Canada for a Cooperative Program Concerning the Development and Procurement of a Space Shuttle Attached Manipulator System. (1974, June). Retrieved during NRC visit 2018, Nov. 8.

7 Munro, M. (1981, April 2). Canada missed chance to send man into space. *The Ottawa Citizen*.

8 Munro, M. (1981, April 2). Canada missed chance to send man into space. *The Ottawa Citizen*.

9 The Canadian Press. (1981, Dec. 11). Quebecers can become Columbia astronauts. *Halifax Mail Star*.

10 Thirsk, R. (2019, March 20). Personal interview.

11 Bolden, C. (2013, Jan. 19). Personal interview.

12 Dotto, L. (1987). *Canada in Space*. Toronto: Irwin. p. 105.

13 Geddes, J. (2017, Jan. 20). How the Trudeau government is bracing for Trump. *Maclean's*. Retrieved from https://www.macleans.ca/politics/ottawa/how-the-trudeau-government-is-bracing-for-trump/.

14 Reagan, R. (1981, March 30). Remarks at the National Conference of the Building and Construction Trades Department, AFL-CIO. *The Reagan Foundation*. Retrieved from https://www.reaganfoundation.org/media/128626/sqp030381.pdf.

15 PBS NewsHour. (2015, March 30). *The day Reagan was shot* [Video]. YouTube. Retrieved from https://www.youtube.com/watch?v=EYI79ziwhow.

16 Brezhnev, L. (1981, March 7). Memorandum for Richard V. Allen:

The White House. Received by the Department of State. *The Reagan Files*. Retrieved from http://www.thereaganfiles .com/19810306.pdf.

17 Reagan, R. (1981, April 24). Letter to President Brezhnev. *The Reagan Files*. Retrieved from http://www.thereaganfiles .com/19810424-2.pdf.

18 Anderson, M., & Anderson, A. (2009). *Reagan's Secret War: The Untold Story of His Fight to Save the World from Nuclear Disaster*. New York: Crown Publishers.

19 Anderson, M., & Anderson, A. (2009). *Reagan's Secret War: The Untold Story of His Fight to Save the World from Nuclear Disaster*. New York: Crown Publishers.

20 Anderson, M., & Anderson, A. (2009). *Reagan's Secret War: The Untold Story of His Fight to Save the World from Nuclear Disaster*. New York: Crown Publishers.

21 Reagan, R. (2007). *The Reagan Diaries* (D. Brinkey, Ed.). New York: HarperCollins Publishers. p. 173.

22 Reagan, R. (2007). *The Reagan Diaries* (D. Brinkey, Ed.). New York: HarperCollins Publishers. p. 203.

23 Reeves, R. (2005). *President Reagan: The Triumph of Imagination*. New York: Simon & Schuster. p. 50.

23 The White House. (1981, May 19). The President's schedule. *The Reagan Library*. Retrieved from https://www.reaganlibrary.gov/sites/ default/files/digitallibrary/smof/president/presidentialbriefingpapers/ box-003/40-439-5730647-003-027-2016.pdf.

25 Benson, C.D., & Compton, W.D. (1983). Years of uncertainty. In *Living and working in space: A history of Skylab*. Scientific and Technical Information Branch, National Aeronautics and Space Administration. Retrieved from https://history.nasa.gov/SP-4208/ ch5.htm.

26 Logsdon, J. (2019, Jan. 14). Personal interview. [Unless otherwise indicated, all Logsdon quotes in this chapter come from that interview.]

27 Canadian Space Agency. (2012, March 1). Mission STS-41G. Retrieved from http://www.asc-csa.gc.ca/eng/missions/sts-041-g .asp. [More thorough details of all of Garneau's experiments are in Lydia Dotto's *Canada in Space*.]

28 Dotto, L. (1987). *Canada in Space*. Toronto: Irwin. p. 28.

29 Remarks by telephone with crewmembers on board the space shuttle Challenger. (1984, Oct. 12). *Reagan Library*. Retrieved from https:// www.reaganlibrary.gov/research/speeches/101284c.

30 Dotto, L. (1987). *Canada in Space*. Toronto: Irwin. pp. 32–33.

31 Woods, W.D., Turhanov, A., & Waugh, L.J. (2017, May 30). Apollo 13: Day 3: 'Houston, we've had a problem.' *Apollo Flight Journal*. Retrieved from https://history.nasa.gov/afj/ap13fj/08day3-problem.html.

32 Woods, W.D., Turhanov, A., & Waugh, L.J. (2017, May 30). Apollo
     13: Day 3: 'Houston, we've had a problem.' *Apollo Flight Journal.*
     Retrieved from https://history.nasa.gov/afj/ap13fj/08day3-
     problem.html.

Chapter Three

1   Brooks, C.G., Grimwood, J.M., & Swenson, L.S. (1979). Moving
    toward operations. In *Chariots for Apollo: A history of manned lunar
    spacecraft.* NASA Special Publication-4205. Retrieved from https://
    www.hq.nasa.gov/office/pao/History/SP-4205/ch8-7.html.
2   Howell, E. (2019, July 19). If Apollo 11 had gone terribly wrong,
    here's what Nixon would have told the country. *Space.com.*
    Retrieved from https://www.space.com/if-apollo-11-astronauts-
    died-nixon-contingency-speech.html.
3   Dotto, L. (1993). *The Astronauts: Canada's Voyageurs in Space.*
    Toronto: Stoddart Publishing Co. p. 53.
4   Tryggvason, B. (2019, Feb. 25). Email to author.
5   Siddon, T. (2019, Jan. 28). Personal interview.
6   Ower, C. (2019, April 17). Personal interview.
7   Truly, R. (2019, Jan. 14). Personal interview.
8   Bolden, C. (2019, Jan. 23). Personal interview.
9   Manley, J. (2019, April 16). Personal interview.
10 O'Keefe, S. (2019, Jan. 15). Personal interview.
11 O'Keefe, S. (2019, Jan. 15). Personal interview.
12 O'Keefe, S. (2019, Jan. 15). Personal interview.
13 Bolden, C. (2019, Jan. 23). Personal interview.

Chapter Four

1   Bondar, R. (2019, March 20). Personal interview. [Unless otherwise
    noted, all of Bondar's quotes in this chapter come from this
    interview.]
2   Gainor, C. (2006). *Canada in Space: The People and Stories Behind
    Canada's Role in the Exploration of Space.* Edmonton: Folklore
    Publishing. pp. 166–167.
3   Gainor, C. (2006). *Canada in Space: The People and Stories Behind
    Canada's Role in the Exploration of Space.* Edmonton: Folklore
    Publishing. pp. 166–167.
4   Howell, E. (2016, Oct. 26). Classified shuttle missions: Secrets in
    space. *Space.com.* Retrieved from https://www.space.com/34522-
    secret-shuttle-missions.html.
5   Cassutt, M. (2009, August). Secret space shuttles. *Air & Space
    Magazine.* Retrieved from https://www.airspacemag.com/space/secret-
    space-shuttles-35318554/?all.

6   Bondar, R. (1994). *Touching the Earth*. Toronto: Key Porter Books. pp. 45–46.

7   Bondar, R. (2019, March 20). Personal interview.

8   Canadian Space Agency. (2018, March 8). Biography of Roberta Lynn Bondar. Retrieved from http://www.asc-csa.gc.ca/eng/astronauts/canadian/former/bio-roberta-bondar.asp.

9   Technically, Robert Stewart was initially recruited, but he resigned one week after the astronaut selection and was swiftly replaced by McKay, an astronaut candidate finalist.

10 Wilson Center. (n.d.). About the Wilson Center. Retrieved from https://www.wilsoncenter.org/about-the-wilson-center.

11 Burke, A., & Jones, R.P. (2020, July 23). Review follows CBC news report that Payette created toxic environment at Rideau Hall. *CBC News*. Retrieved from https://www.cbc.ca/news/politics/pco-independent-review-governor-general-1.5661236?cmp=rss.

12 Tunney, C. (2019, July 3). Gov. Gen. Payette won't be moving into Rideau Hall for now. *CBC News*. Retrieved from https://www.cbc.ca/news/politics/payette-rideau-hall-move-1.5192368.

13 Donovan, K. (2017, Aug. 21). Incoming governor general Julie Payette drops fight to keep divorce records sealed. *The Toronto Star*. Retrieved from https://www.thestar.com/news/canada/2017/08/21/incoming-governor-general-julie-payette-drops-fight-to-keep-divorce-records-sealed.html.

14 NASA. (2007, Aug. 29). NASA safety review finds no evidence of improper alcohol use by astronauts before space flight. Retrieved from https://www.nasa.gov/home/hqnews/2007/aug/HQ_07184_oconnor_alcohol_study.html.

Chapter Five

1   Hadfield, Chris. Personal interview.

2   Evans, Mac. Personal interview.

3   Howell, E. (2018, April 24). 15 space travel tips from an astronaut. *Space.com*. Retrieved from https://www.space.com/40386-15-astronaut-space-travel-tips.html.

4   Bolden, C. (2019, Jan. 23). Personal interview.

5   Wright, R. (Ed.). International Space Station Program Oral History Project: Edited Oral History Transcript, Melanie Saunders. NASA Johnson Space Center. Retrieved from https://historycollection.jsc.nasa.gov/JSCHistoryPortal/history/oral_histories/ISS/SaundersM/SaundersM_8-6-15.htm.

6   Zak, A. (2019, April 10). The Zvezda Service Module, SM. *Russian Space Web*. Retrieved from http://www.russianspaceweb.com/iss_sm.html.

7 Leary, W.E. (2000, July 11). Space station waits upon the launch of new "star." *New York Times.* Retrieved from https://archive. nytimes.com/www.nytimes.com/learning/teachers/featured_ articles/20000711tuesday.html.

8 Parazynski, S. (2019, April 3). Personal interview.

9 Thirsk, R. (2019, Jan. 10). Personal interview.

10 Williams, D. (2019, Jan. 29). Personal interview.

11 Rakobowchuk, P. (2013, Feb. 1). Canada space agency boss leaves, with long-term federal plans for space unclear. *Maclean's.* Retrieved from https://www.macleans.ca/general/canada-space-agency-boss-leaves-with-long-term-federal-plans-for-space-unclear/.

12 Emerson, D. (2019, Feb. 4). Personal interview.

13 Natynczyk, W. (2019, Feb. 26). Personal interview.

14 Parazynski, S. (2019, April 3). Personal interview.

15 Parazynski, S. (2019, April 3). Personal interview.

16 Podwalski, K. (2019, Feb. 19). Personal interview.

17 Parazynski, S. (2016, Sept. 9). Gr8 to see @Astro_Kate7 doing a super job out there! And yes, our repairs are still under warranty . . . @Astro_Wheels [Tweet]. Twitter. Retrieved from https://twitter.com/AstroDocScott/status/774359253964300288.

Chapter Six

1 Dodge, Michael. Personal interview.

2 Fisher, M. (2014, Sept. 3). Everything you need to know about the Ukraine crisis. *Vox.* Retrieved from https://www.vox.com/2014/9/3/18088560/ukraine-everything-you-need-to-know.

3 Zak, A. (2015, Sept.). A rare look at the Russian side of the space station: How the other half lives. *Air & Space Magazine.* Retrieved from https://www.airspacemag.com/space/rare-look-russian-side-space-station-180956244/.

4 Thirsk, Bob. Personal interview.

5 Special thanks to NASA's website for this helpful diagram: https://www.nasa.gov/sites/default/files/thumbnails/image/iss_config_exploded_view_page_0.jpg.

Chapter Seven

1 Kutryk, J. (2019, Jan. 22). Personal interview.

2 Pearlman, R. (2018, Aug. 27). NASA astronaut candidate resigns prior to qualifying for spaceflight. *collectSPACE.* Retrieved from http://www.collectspace.com/news/news-082718a-kulin-nasa-astronaut-candidate-resigns.html.

3 Howell, E. (2018, Sept. 13). Meet the astronaut "den mother" who takes care of NASA's 2017 astronaut candidate class. *Space.com.*

Retrieved from https://www.space.com/41783-astronaut-den-mother-jeremy-hansen-2017-class.html.

4   Glover, V. (2019, April 1). Personal interview.

5   Hansen, J. (2019, June 3). Personal interview.

6   Canadian Space Agency. (2013, Feb. 4). Biography of Steve MacLean. Retrieved from http://www.asc-csa.gc.ca/eng/astronauts/canadian/former/bio-steve-maclean.asp.

7   Brown, C. (2018, Nov. 25). Governor General to attend space launch for Canadian astronaut David Saint-Jacques. *CBC News*. Retrieved from https://www.cbc.ca/news/technology/david-saint-jacques-canadian-astronaut-iss-julie-payette-1.4919882.

8   Kutryk, J. (2019, Jan. 22). Personal interview.

9   Podwalski, K. (2019, Feb. 19). Personal interview.

10 Shireman, K. (2019, March 13). Personal interview.

11 Space.com Staff. (2013, Feb. 7). 'Star Trek' actors beam hellos to astronaut in space (Photos). *Space.com*. Retrieved from https://www.space.com/19673-star-trek-chris-hadfield-william-shatner-photos.html.

12 Howell, E. (2013, May 15). Canada celebrates star astronaut Chris Hadfield's return to Earth. *Space.com*. Retrieved from https://www.space.com/21149-canada-astronaut-chris-hadfield-party.html.

13 CTV.ca News Staff. (2008, June 1). Montreal-born astronaut brings bagels into space. *CTV News*. Retrieved from https://www.ctvnews.ca/montreal-born-astronaut-brings-bagels-into-space-1.299619.

14 NASA. (2018, March 28). Preflight interview: Gregory Chamitoff. Retrieved from www.nasa.gov/mission_pages/station/expeditions/expedition17/exp17_interview_Chamitoff.html.

15 NASA. (2011, July). Biographical data: Gregory Errol Chamitoff (Ph.D.). Retrieved from https://www.nasa.gov/sites/default/files/atoms/files/chamitoff_gregory.pdf.

16 Feustel, D. (2019, March 5). Personal interview.

17 NASA. (n.d.). Andrew J. Feustel (PhD) NASA Astronaut. Retrieved from https://www.nasa.gov/astronauts/biographies/andrew-j-feustel/biography.

18 Feustel, D. (2019, March 5). Personal interview.

19 @IndiraFeustel. (2019, June 27). Here's a question for all of you. Should @astro_feustel wash the space smell off his spacesuit and it be gone forever or wear it to his next PR event, as is. @nasa @NASA_Astronauts #Expedition56 [Tweet]. Twitter. Retrieved from https://twitter.com/IndiraFeustel/status/1144093073342894080.

Chapter Eight

1   Fife, R., Chase, S., & Fine, S. (2019, Feb. 7). PMO pressed Wilson-Raybould to abandon prosecution of SNC Lavalin. *The Globe and Mail*. Retrieved from https://www.theglobeandmail.com/politics/

article-pmo-pressed-justice-minister-to-abandon-prosecution-of-snc-lavalin/.

2 Ogilvie, K. (2019, Feb. 28). Personal interview.

3 Howell, E. (2019, March 1). Canada joins NASA's Lunar Gateway station project with 'Canadarm3' robotic arm. *Space.com.* Retrieved from https://www.space.com/nasa-lunar-gateway-canada-canadarm3-robot-arm.html.

4 CityNews Toronto. (2019, Feb. 28). Trudeau is now answering questions from the media, the first is about the testimony of Jody Wilson-Raybould [Tweet]. Twitter. Retrieved from https://twitter.com/CityNews/status/1101140268630323201.

5 Bruemmer, R. (2019, Feb. 28). PM Trudeau reflecting on Wilson-Raybould's presence in Liberal caucus. *The Montreal Gazette.* Retrieved from https://montrealgazette.com/news/local-news/pm-trudeau-reflecting-on-wilson-rayboulds-presence-in-liberal-caucus.

6 Semeniuk, I. (2019, Feb. 28). Canada's space program sets new course with historic commitment to lunar outpost. *The Globe and Mail.* Retrieved from https://www.theglobeandmail.com/canada/article-canada-is-going-to-the-moon-trudeau-announces-canada-will-join-the/.

7 Mortillaro, N. (2019, Feb. 28). Canada's heading to the moon: A look at the Lunar Gateway. *CBC News.* Retrieved from https://www.cbc.ca/news/technology/canada-lunar-gateway-1.5037522.

8 Lambright, W.H. (2009). Launching a new mission: Michael Griffin and NASA's return to the moon. IBM Center for the Business of Government.

9 Beeby, D. (2019, April 28). New US moon-landing timetable throws Canada's Lunar Gateway role into question. *CBC News.* Retrieved from https://www.cbc.ca/news/politics/space-canadarm-nasa-lunar-gateway-1.5112559.

10 Financial Post Business Wire. (2019, Aug. 19). MDA selected to build robotic interfaces for Canadarm3 on NASA-led Gateway. *Financial Post.* Retrieved from https://business.financialpost.com/pmn/press-releases-pmn/business-wire-news-releases-pmn/mda-selected-to-build-robotic-interfaces-for-canadarm3-on-nasa-led-gateway.

11 Government of Canada. (2019). *Exploration, imagination, innovation: A new space strategy for Canada.* Retrieved from http://www.asc-csa.gc.ca/pdf/eng/publications/space-strategy-for-canada.pdf.

12 Boucher, M. (2019, June 27). Don't Let Go Canada and changes at the Canadian Space Agency resulted in government funding commitment. *SpaceQ.* Retrieved from http://spaceq.ca/dont-let-go-canada-and-changes-at-the-canadian-space-agency-resulted-in-government-funding-commitment/.

13 Boucher, M. (2019, June 27). Don't Let Go Canada and changes at the Canadian Space Agency resulted in government funding commitment. *SpaceQ.* Retrieved from http://spaceq.ca/dont-let-go-canada-and-changes-at-the-canadian-space-agency-resulted-in-government-funding-commitment/.

14 Nadeau, C. (2010). *Rogue in Power: Why Stephen Harper is Remaking Canada by Stealth.* Toronto: James Lorimer & Co. Ltd., Publishers. pp. 87–88.

15 Turner, C. (2013). *The War on Science: Muzzled Scientists and Wilful Blindness in Stephen Harper's Canada.* Vancouver: Greystone Books.

16 The Canadian Science Writers' Association was one of the organizations that were very outspoken about muzzling during Harper's tenure and advocating for change in the public sphere. I was a member of the organization at the time, and today I am president of the successor group, Science Writers and Communicators of Canada. However, I did not participate directly in the Harper campaign.

17 Zussman, D. (2016). Stephen Harper and the Federal Public Service: An uneasy and unresolved relationship. In J. Ditchburn, & G. Fox. (Eds.), *The Harper Factor: Assessing a Prime Minister's Policy Legacy.* Montreal and Kingston: McGill-Queen's University Press.

18 Office of the Conflict of Interest and Ethics Commissioner. (2019, Aug. 14). Contravention of section 9 of the Conflict of Interest Act found in report released by Commissioner Dion. Retrieved from https://ciec-ccie.parl.gc.ca/en/news-nouvelles/Pages/NR08142019.aspx.

19 Tasker, J.P. (2019, Aug. 14). 'I take responsibility,' Trudeau says in wake of damning report on SNC-Lavalin ethics violation. *CBC News.* Retrieved from https://www.cbc.ca/news/politics/trudeau-snc-ethics-commissioner-violated-code-1.5246551.

20 Laporte, S. (2019, May 6). Personal interview.

21 Howell, E. (2019, Aug. 14). How big pharma was wooed to space-based 'business park.' *Forbes.* Retrieved from https://www.forbes.com/sites/elizabethhowell1/2019/08/14/how-big-pharma-was-wooed-to-space-based-business-park/#10224e2a32e1.

Epilogue

1   As of April 2020, McClain's estranged spouse is accused of "making false statements to federal authorities" concerning an accusation made against McClain, but nothing has been proven in court concerning either of the two women. More details in this Space.com article: https://www.space.com/astronaut-anne-mcclain-wife-charged-lying-space-crime.html.

2   Canadian Space Agency. (2019, Feb. 11). *A memorable meeting between Jenni Sidey-Gibbons and Roberta Bondar*

[Video]. YouTube. Retrieved from https://www.youtube.com/
watch?v=xf7CUfLOjrE.

3   Canadian Space Agency. (2019, June 24). *LIVE — Return to Earth of
    David Saint-Jacques and his crewmates* [Video]. YouTube. Retrieved
    from https://www.youtube.com/watch?v=gUFUS3Mysxg.

4   CBC News. (2019, June 24). 'Gravity isn't my friend,' Canadian
    astronaut David Saint-Jacques says as he and his crewmates return to
    Earth. Retrieved from https://www.cbc.ca/news/technology/david-
    saint-jacques-1.5187999.

5   Shireman, K. (2019, March 13). Personal interview.

6   Hopkins, M. (2019, April 1). Personal interview.

7   At the time I did the interview, Glover and Hopkins were the only
    people assigned to the mission. Walker and Noguchi were added in
    March 2020. At least one media report said the Russians were invited
    to join but refused to fly this demonstration mission since the
    spacecraft is still in the testing phase: https://spacenews.com/nasa-
    selects-astronauts-for-crew-dragon-mission/. There is no mention
    of Canada being invited, but presumably our country is still out of
    flight credits as Saint-Jacques only completed his mission in 2019,
    and the next Canadian mission would be later in the 2020s.